FIELD GUIDE TO THE

SĀMOAN ARCHIPELAGO

Fish, Wildlife, and Protected Areas

Written and Illustrated by
Meryl Rose Goldin

THE
BESS
PRESS

3565 Harding Ave. Honolulu, Hawai'i 96816
www.besspress.com

To my husband, Christopher E. Stein, superintendent of the National Park of American Sāmoa (during the time this book was researched, written, and illustrated), who gave me the greatest gift of all: the chance to live in the enchanted isles of American Sāmoa and snorkel to my heart's content

Design: Carol Colbath
Index: Lee S. Motteler

Library of Congress Cataloging-in-Publication Data

Goldin, Meryl Rose
  Field guide to the Samoan archipelago :
fish, wildlife, and protected areas / written
and illustrated by Meryl Rose Goldin
      p. cm.
  Includes bibliography, index,
illustrations.
  ISBN 1-57306-111-5
  1. Marine biology - Samoan islands.
  2. Island ecology - Samoan islands.
  3. Natural history - Samoan islands.
  4. Marine parks and reserves - Samoan islands.
  5. Wildlife conservation - Samoan islands.
  I. Title
QH198.S3.G67 2002   574.92-dc20

Printed in Korea

This publication has been funded in part by the following organizations:

United States Department of Interior, Office of Insular Affairs

American Sāmoa Department of Marine and Wildlife Resources

American Sāmoa Environmental Protection Agency

American Sāmoa Community College Land Grant Program

Bat Conservation International, Inc.

American Sāmoa Department of Commerce, Coastal Zone Management Program

Fagatele Bay National Marine Sanctuary

Sāmoa Department of Agriculture, Fisheries and Forests

National Park Service

Sea Grant Program of the University of Hawai'i

Rotary Club of American Sāmoa

U.S. Fish and Wildlife Service, Region 1

South Pacific Regional Environmental Programme (SPREP) and the South Pacific Biodiversity Conservation Programme

This publication is funded in part by GEN 103, U.S. Department of Interior Office of Insular Affairs. The views expressed herein are those of the author and do not necessarily reflect the views of DOI, or any of its subagencies.

Although the information in this document has been funded in part by the U.S. Environmental Protection Agency under assistance agreement NE-999807-01-0 to the American Sāmoa Community College Land Grant Program, it may not necessarily reflect the views of the Agency and no official endorsement should be inferred.

CONTENTS

FOREWORD

The tropical islands of the Sāmoan Archipelago, including both the U.S. Territory of American Sāmoa and the nearby independent nation of Sāmoa, are some of the most enchanting islands in the world. In April 1997, I delighted in visiting these islands for American Sāmoa's Flag Day ceremony and to dedicate one of the United States' newest national parks, the National Park of American Sāmoa. As the plane circled over the main island of Tutuila, I looked down in wonder upon this beautiful emerald isle and the turquoise waters of its fringing reef seemingly floating in the deep, cobalt blue of the Pacific Ocean. One must use superlatives to describe this enchanting place. During my visit I snorkeled on the reef of the American Sāmoa island of Ofu, one of the sacred Manuʻa Islands, the legendary birthplace of Polynesia. The Ofu reef, along with many other reef areas encircling all islands comprising the Archipelago, is one of the most pristine coral reefs in the South Pacific.

In the *Field Guide to the Sāmoan Archipelago*, written and illustrated by Meryl Rose Goldin, the people of Sāmoa and elsewhere will learn about the wondrous diversity of life found throughout the Sāmoan Archipelago. Almost one thousand species of coral reef fish swim among the hundreds of species of coral, flying fox fruit bats grace the skies, and colorful forest and sea birds nest throughout the tangled branches of the rain forest trees and upon gleaming sand beaches.

This field guide will assist readers in exploring the islands and its fish and wildlife. With its help, they can discover that the Sāmoan Islands have much to show those who are looking for the natural wonders of our Earth, and much worth saving for future generations.

Bruce Babbitt, former
U.S. Secretary of the Interior

ACKNOWLEDGMENTS

This book is respectfully written for the people of the Sāmoan Archipelago. Its completion would not have been possible without the steadfast project support of Lydia Faleafine-Nomura and the Office of Insular Affairs, United States Department of Interior, and Tumarii Tutangata and Iosefatu Reti of the South Pacific Regional Environmental Programme.

I would also like to take this opportunity to acknowledge, in memoriam, Brooks Harper, United States Fish and Wildlife Service, who spent many years working to protect the fish and wildlife of the entire Pacific region.

My colleagues in the archipelago helped in many ways, but first and foremost, I must acknowledge my dear friend, creative advisor, snorkeling partner, and fellow biologist Mary Hake.

I am indebted to other friends, colleagues, and specialists who assisted with general reviews, contributed information or literature, and resolved specific problems, including: Don Barclay (nudibranchs), Charles E. Birkeland (coral reefs, crown-of-thorns starfish), Anne P. Brooke (bats), Steve Brown (western islands of Sāmoa), Robert P. Cook (lepidoptera), Robert H. Cowie (land snails), Alison Green (fishes), Elizabeth Flint (seabirds, Rose Atoll), Holly Freifeld (birds), Onolina Fuamatu (Sāmoan nomenclature), Cindy Hunter (corals), Stephen E. Miller (land snails), Michael Molina (Rose Atoll), H. Douglas Pratt (birds), Joshua Seamon (birds), Barry D. Smith (macroinvertebrates), Stan Sorenson (herpetofauna), Funealii Lumaava Sooaemalelagi (western islands of Sāmoa), Rowland R. Shelley (millipedes), Epifania Suafoa (Sāmoan nomenclature), Sabina Fajardo Swift (millipedes), Tavita Togia (Sāmoan nomenclature), Ailao Tualaulelei (birds), Merlin Tuttle (bats), Ruth Utzurrum (bats), and, last but not least, W. Arthur Whistler (rain forest, plants).

Other esteemed individuals scattered throughout the islands of the Sāmoan Archipelago and elsewhere helped in many ways and I would like to express my sincere appreciation to them for personally contributing time, knowledge, and technical resources that made the production of this field guide possible: Fale Aitatoto, Leota Leuluaialiʻi, Alapapa Vaea Ainuʻu, Fini Aitaoto, Dion Ale, Sao Ala, James Atherton, T. D. Bear, Steve Brown, Jeff Burgett, Robert Cook, Nafanua Paul Cox, Peter Craig, Chuck Cranfield, Darryl Cunningham, Nancy Daschbach, Christine Willis, Thomas Elmqvist, John Enright, Carl Goldstein, Siafoi Faaumu, Taufeteʻe John Faumuina, Mino Fialua, Mary Hake, Rita Hunkin, Sonja Siainiusami Hunter, Cindy Hunter, Wallace H. Jennings, Dale and Adeline Pritchard Jones, Punipuao Lagai-Nagalapadi, Asila Philip Langford, Sarah Lendon, Larry Madrigal, Tito and Marge Malae, Laʻavasa Malua, Amby Mara, Francois Martel, Leslie McEwen, Joanne Meader, David Meader, Sue Miller, Michael Molina, Charlotte Orville, Denise Rall, Verne and Marion Read, Iosefatu Joe Reti, Sarah Risi, Andy Risi, Barry I. Rose, Suesan Saucerman, Seumanutafa Aeau Tiavalo Seumanutafa, Chris Solek, Tina Sooto, Christopher Stein, Tanielu Sua, Saʻopapa Taifane, Dorinda Talbot (and the *Lonely Planet Guide to Samoa: Independent and American*), Muaʻavatele Togipa Tausaga, Aiono Alofa Tuaumu, Malelega Tuiolosega, Ufagafa Ray Tulafono, Tamariʻi Tutangata, Carol Umebayashi, Don Vargo, Walter Vermeulen, Sheila Wiegman, David Worthington, and Patricia Young.

Many persons and organizations provided technical or administrative assistance, some thanked elsewhere: Governor Tauese Sunia, Lieutenant Governor Togiola Tulafono, and Galumalemana Frank Pritchard, American Sāmoa Government; Congressman Eni Faleomavaega; staff of the United States Department of Interior, including Bruce Babbitt, Al Stayman, Danny Aranza, Nikalao Pula, Rosemarie Babel, William Brown; staff of the National Park Service, National Park of American Sāmoa; Ufagafa Ray Tulafono and staff of the Department of Marine and Wildlife Resources, American Sāmoa Government; Tamariʻi Tutangata, Iosefatu Reti, Neva Wendt, Sue Miller, and staff of the South Pacific Regional Environmental Programme (SPREP) and the South Pacific Biodiversity Conservation Programme; Taufete'e John Faumuina, LeLei Peau, Nancy Daschbach, and staff of the Coastal Zone Management Program of the American Sāmoa Department of Commerce; Mua'avatele Togipa Tausaga and Sheila Wiegman of the American Sāmoa Environmental Protection Agency; Norm Lovelace, Patricia Young, and Carl Goldstein of the United States Environmental Protection Agency; Don Vargo, Carol Whittaker, and staff of American Sāmoa Community College Land Grant Program; Andy Risi and Jennifer Aicher of the American Sāmoa Community College Marine Science Program; Seumanutafa Aeau Tiavalo Seumanutafa, Pita Liu, and Easter Galuvao of (Western) Sāmoa Department of Lands, Surveys and Environment; Sonja Siainiusami Hunter, Alise Faulolo-Stunnenberg, and the (Western) Sāmoa Visitors Bureau; Joanne Meader and the Australia High Commission; Ecotour Samoa; Walter Vermeulen and Dion Ale of O le Siosiomaga Society, Inc.; Verne and Marion Reade through Bat Conservation International; Jim Porter and Connie Porter of Samoa Air; and John Enright of Le Vaomatua.

I would like to express my gratitude to the following organizations for providing monetary support for production costs associated with publication: United States Department of Interior, Office of Insular Affairs; American Sāmoa Department of Marine and Wildlife Resources; South Pacific Regional Environmental Programme (SPREP) and the South Pacific Biodiversity Conservation Programme; United States Fish and Wildlife Service, Region 1; Coastal Zone Management Program and Fagatele Bay National Marine Sanctuary of the American Sāmoa Department of Commerce; United States Environmental Protection Agency, Region 9; American Sāmoa Environmental Protection Agency; American Sāmoa Community College Land Grant Program; Bat Conservation International; Sea Grant Program of the University of Hawai'i; Seacology Foundation;

Samoa Air; Le Vaomatua, Inc., in memoriam to Iris Brooks; Rotary Club of American Sāmoa; AUSAid; and Sāmoa Ministry of Agriculture, Fisheries, and Forests.

I would also like to thank the following authors, publishers, and publishing houses for guidance or for use of reference photographs for my watercolor illustrations (see references for detailed information on each publication): John Randall, Gerald R. Allen, Roger Steene, Tony Crawford, and Crawford House Press, *Fishes of the Great Barrier Reef and Coral Sea*; Charles Arneson and Patrick L. Colin, *Tropical Pacific Invertebrates*; William Hamilton and University of Hawai'i Press; Sean McGowan, *Reptiles and Amphibians of the Hawaiian Islands*; Robert Myers, *Micronesian Reef Fishes*; and H. Douglas Pratt, *A Field Guide to the Birds of Hawaii and the Tropical Pacific*.

Thank you to my publisher, The Bess Press, and Buddy Bess (publisher), Revé Shapard (editor), and Carol Colbath (layout and design) who were superbly supportive and creative, making publishing this book a reality and lots of fun. A big thank you to the author friends in my life who provided the inspiration needed to endeavor my first book: Virginia Carpenter Brown, Nancy Bohac Flood, Nardia Rose Goldin, Elizabeth A. Lawrence, and Robert W. Nero.

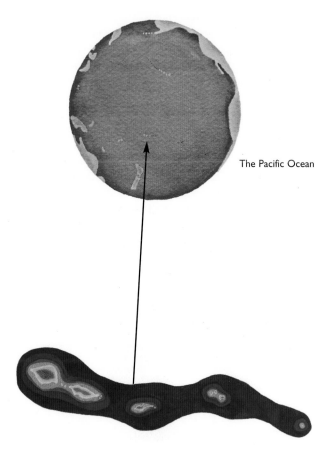

The Pacific Ocean

The Sāmoan Archipelago

INTRODUCTION

The South Pacific is a beautiful and enchanting part of the planet, and the Sāmoan Archipelago exemplifies that beauty and magic in its coral reef and rain forest habitats. Few other habitats in the world exhibit nature and its creatures in such vivid shades of green, turquoise, blue, pink, yellow, orange, purple, red, and brown.

Field Guide to the Sāmoan Archipelago: Fish, Wildlife, and Protected Areas has been written to enhance the Sāmoa Islands' overall natural resource management program, to heighten public awareness of and support for ecosystem protection and stewardship, and to further environmental education in local schools. It features over 500 illustrations of the species of marine and terrestrial fauna that constitute the wealth of biodiversity found in the ecosystems of Sāmoa's coral reefs, ocean waters, sea grass beds, rain forests, mangrove forests, and wetlands. It also describes the "protected areas" (parks, reserves, and biodiversity conservation sites) and the natural history of the wildlife, using species-species and species-environment interactions (including our own human foibles) to highlight the critical environmental issues facing the archipelago, such as exploitation of forest and marine resources, nonpoint source pollution, and sedimentation of the reefs.

The *Field Guide* has been written to serve several purposes for several audiences: as an informative guide for tourists and other visitors to the archipelago, as a readable reference for anyone interested in the natural history and biodiversity of the Sāmoa Islands, and as a supplementary biology or environmental education text for Sāmoan students.

SCOPE

The wildlife species covered in this book are, for the most part, the most frequently encountered species of coral reef and nearshore fishes and marine invertebrates, as well as the commonly observed mammals, birds, and reptiles, the common and endangered land snails, and a few frequently encountered insects and arthropods. I have also included tables of birds, reptiles, and amphibians that contain names of all species reported in the archipelago. The phrase "reported or recorded for the archipelago," does not mean that a species occurs on every island in the archipelago, but rather has been recorded for one or more islands in both island nations, American Sāmoa and Sāmoa (see discussion of independent "Western" Sāmoa's name change below).

Species chosen for illustration were those listed or highlighted in published and unpublished data and reports, species accounts, and scientific literature compiled by both visiting scientists and staff of various government agencies in both Sāmoas. Further, I included most endemic, endangered, rare, noteworthy, and nonnative wildlife species. Endemic (and restricted) secies are listed as such. In almost all instances, the marine and terrestrial species illustrated can be found in their appropriate habitats throughout the archipelago, not just within protected areas. References used to determine a species' rarity were the *U.S. Endangered Species List* and *2000 IUCN Red List of Threatened Animals*. The World Conservation Union–IUCN (International Union for the Conservation of Nature and Natural Resources) is a global conservation organization, comprising scientists from all over the world, that tracks the health and welfare of all living species on earth. Its *Red List of Threatened Animals* (now combined with its list of plants in the *2000 Red List of Threatened Species*™*)* is a late-warning database of rare and endangered species that uses strict scientific criteria to define the status of rare and endangered species. Unfortunately, several species from the archipelago are rare enough to be listed in this database as well as on the U.S. Endangered Species List.

As defined here, the Sāmoan Archipelago comprises the

seven main islands of American Sāmoa and the four main islands of the newly named independent nation of Sāmoa (formerly Western Sāmoa). To eliminate confusion, I use the term "western islands of Sāmoa" or "independent Sāmoa" to refer to the former Western Sāmoa and, following the U.S. Board on Geographic Names, "Sāmoa Islands" (rather than "Sāmoan Islands") to refer to the entire archipelago. The text is interspersed with insets to guide the reader in thinking about the environmental challenges facing the Sāmoa Islands and about what can be done to protect their magnificent biological diversity.

The metric system is used throughout, with English/U.S. equivalents in parentheses. Taxonomy and best-known English names of birds follow widely used books describing the avifauna of the area in general, including Pratt et al. (1987); seabirds follow Harrison (1983 and 1987); and shorebirds follow Hayman et al. (1986). Additional information, including lesser-known but accepted names of forest and terrestrial birds, follows Clements (1991). Names of fishes follow Randall et al. (1990); marine invertebrates follow Colin and Arneson (1995); and herpetofauna follow Zug (1991) and McKeowen (1996). Names of butterflies (lepidoptera) follow Kami and Miller (1998), and land snails, Cowie (1998). Maximum length of fish species is given in parentheses following the scientific name; body length (and wingspan where available) for birds follows the scientific name. Sāmoan nomenclature is provided in boldface type. Most distances are from University of Hawai'i Cartographic Laboratory (1981). Distances, as well as other measures, may be approximate.) The following abbreviations from the IUCN *Red List* are used to indicate applicable threatened species categories:

CR   Critically endangered; facing an extremely high risk of extinction in the wild in the immediate future

EN   Endangered; facing a very high risk of extinction in
      the wild in the near future
VU   Vulnerable; facing a high risk of extinction in the wild
      in the medium-term future
DD   Data deficient; though species thought to be in decline,
      inadequate information exists to make an assessment
      of risk

GEOGRAPHY

The Sāmoan Archipelago (**atumotu o Sāmoa**) is a
chain of volcanic islands stretching for more than 480 km
(298 mi.) in an east-west orientation between 13 and 15
degrees south of the equator, longitude 168–172 degrees.
(Swains Island is outside this area, at 11 degrees south.) The
archipelago is "just east of tomorrow" (the International
Dateline), approximately 4,184 km (2,600 mi.) southwest of
Hawai'i, and 2,900 km (1,800 mi.) northeast of New
Zealand (fig. 1).

The archipelago comprises nine inhabited islands
(**motu**) and many uninhabited islets, plus two distant coral
atolls (**motu tu'ufua**), Swains Island (inhabited) and Rose
Atoll (uninhabited). It is divided politically into American
Sāmoa (five high islands and two atolls) and the western
islands of Sāmoa (four islands) (fig. 2). American Sāmoa is
an unincorporated trust territory of the United States and
includes the islands of Tutuila, 'Aunu'u, the Manu'a Islands
(Ofu, Olosega, and Ta'u), plus both coral atolls (Rose Atoll
and Swains Island). The newly named, independent nation
of Sāmoa consists of two large islands, 'Upolu and Savai'i,
and two smaller islands, Apolima and Manono, which are
situated between the two larger islands. Many other tiny
islands (islets) occur offshore from the main islands of the
archipelago. The two nations are separated by a strait, 64 km
(just under 40 mi.) wide.

American Sāmoa and the western islands of Sāmoa

share the same language and culture, but have unique land-
scape features. In addition, several species are endemic to,
i.e., occur only on, single islands within the archipelago. All
of the islands are volcanic in origin.

American Sāmoa
Tutuila and ‘Aunu‘u

Tutuila is the largest island of American Sāmoa (140
sq. km/54 sq. mi.) and comprises the remnants of seven
major volcanoes, two of which are now submerged at either
end of the island. The five others are part of the now visible
mountain groups. The highest mountain on Tutuila is
Matafao, at 653 m (2,142 ft.). A barrier reef that once encircled Tutuila is now submerged at approximately the 100-
fathom line (600-ft. contour line) offshore. It was drowned
by subsidence of the island coupled with rising sea level at
the end of the Ice Age. The tiny island of ‘Aunu‘u (approximately 2 sq. km/.77sq. mi.) is located off the eastern end of
Tutuila.

Manu‘a Island Group

Approximately 102 km (63 mi.) east of Tutuila lies the
Manu‘a group, comprising the islands of Ofu, Olosega, and
Ta‘u. The two small islands of Ofu and Olosega total only 9
sq. km in area (3.5 sq. mi.), and their highest points are
Tumutumu Mountain, at 494 m (1,621 ft.), and Piumafua
Mountain, at 639 m (2,095 ft.) respectively. Ta‘u is 39 sq.
km in area (15 sq. mi.), with the highest point in all of
American Sāmoa, Mount Lata, at approximately 966 m
(3,170 ft.).

Rose Atoll (Nu‘u o Manu)

Farthest to the east is Rose Atoll, one of the smallest
atolls in the world. This 4-sq.-km (1.5-sq.-mi.) atoll, a U.S.
National Wildlife Refuge, lies approximately 257 km (160

mi.) east of Tutuila and 156 km (97 mi.) east of the Manu'a group. Two tiny islets, Rose and Sand, sit upon the edge of this rectangular atoll. Its central lagoon is only 15 m (49 ft.) deep, and its floor is dotted with tall coral pinnacles that rise to the ocean's surface. There is no access to this atoll without a permit. For more information see Protected Reserves chapter.

Swains Island

Swains Island, located approximately 370 km (230 mi.) north of the archipelago, is politically part of American Sāmoa, but geographically part of the Tokelau archipelago. Swains Island is actually a coral atoll, but unlike Rose Atoll, its central lagoon has no connection to the sea and its water is brackish (part salt, part fresh). Though Swains Island is part of Polynesia, many fish species common there are more characteristic of the marine fauna of Micronesia and are not known to occur elsewhere in the archipelago (Green 1996).

The Western Islands of Sāmoa
'Upolu

'Upolu is the second largest island in the archipelago, comprising 1,115 sq. km (430 sq. mi.). It has the most waterfalls, valleys, and rivers of all the islands. Its mountains rise to 1,115 m (3,658 ft.).

Savai'i

Savai'i, 18 km (11 miles) west of 'Upolu, is the largest, but one of the least densely populated islands in the archipelago, comprising 1,700 sq. km (656 sq. mi.). Mt. Silisili is the highest mountain in the archipelago, towering 1,857 m (6,095 ft.) in the center of the island. Savai'i is the site of the archipelago's most recent volcanism, which occurred between 1902 and 1911.

Manono, Apolima, and the Aleipata Group

The islands of Manono and Apolima occur in the channel between the two larger islands of ʻUpolu and Savaiʻi. Several other tiny islands and islets are located offshore. Especially notable are the Aleipata group (Nuʻutele and Nuʻulua islands) off the eastern shore of ʻUpolu; not only are they visually beautiful, but they also provide important habitat for seabirds.

CLIMATE

The average annual temperature in the Sāmoa Islands is about 27.5 degrees C (81.5 degrees F). The southern hemisphere's winter months of June, July, and August are the coolest, while the summer months of January, February, and March are the warmest. "Winter" is only a relative term, and the annual range in Sāmoa's temperature is only about 2 degrees C. The prevailing tradewinds are southeasterly and are most constant during the austral winter months.

All of the Sāmoa Islands are tropical and therefore receive a lot of rain (**timu**). Rainfall averages 318 cm (125 in.) annually at the Pago Pago International Airport, but varies greatly over the islands because of their dramatic topography. For example, Pago Pago Harbor receives approximately 510 cm (200 in.) annually, while Tutuila's mountains receive approximately 750 cm (almost 300 in.) annually. The cloud forest of Mt. Lata on the island of Taʻu is thought to receive over 1,000 cm (400 in.). Apia, the capital of the western islands of Sāmoa, receives only about 290 cm (114 in.) of rain annually. During the rainy season (**vaitaimi o timuga**), October through March, relative humidity (**sau o le ea**) averages above 80%; about half of all annual rainfall usually occurs December through March. Humidity averages 70% during the relatively drier months (**vaitaimi o eleʻele matūtū).**

Hurricanes (**afā**) that strike the Sāmoan Archipelago

usually originate from the north, but occasionally from the east or west. Historically, Sāmoa experienced 39 hurricanes between 1831 and 1926, with most (85%) occurring from December through March. While hurricanes are sporadic, several destructive ones hit the islands recently. Tusi (1987), Ofa (1990), and Val (1991) hit different areas within the archipelago, and the islands and surrounding reefs incurred varying amounts of damage ranging from minimal to extreme.

SNORKELING AND DIVING SAFETY AND ETIQUETTE

Le Ava: Reef Channels

It is important to learn about **le ava,** a feature of reefs, before venturing into unknown areas. Much of the incoming water associated with waves must flow back off the reef somewhere. Channels in the reef, connecting the lagoon or reef flat with the outer reef slope and ocean, provide a return passage for the incoming water. Such channels are known as **le ava** in the Sāmoa Islands. Deep channels occurring in reefs may contain a high diversity of fishes and corals. Nevertheless, snorkelers and swimmers should avoid the channels, which typically have an extremely strong current flowing outward. In addition, there may be strong currents parallel to the shoreline flowing toward **le ava**.

Le ava can usually be located from an elevated area on shore. Look for an area of darker (deep) water extending from near the shore out through the wave zone. You will often notice **le ava**'s rippling water as it flows out the channel against the water of incoming waves. The current through **le ava** is usually strongest from mid- to high tide, or when waves are big, and weakest at low tide, or when the sea is calm.

One additional word of caution: Rocky coastlines can

be hazardous. Unexpected large waves, sometimes called "phantom waves," have been known to wash the unwary, or those who venture too close to the water's edge, into the sea.

Safety Precautions for Snorkeling and Diving

• Wear plenty of sunscreen.

• Get information on a chosen area, including current weather conditions and surf reports, before snorkeling or diving there.

• Never snorkel in high-surf conditions. The best time to snorkel is usually during "slack tide" (one hour on either side of low or high tide), when currents are apt to be at a minimum. Low tide is best for a beginner, but some reefs are too shallow during low tide for optimal snorkeling.

• Find the best snorkel spots for your ability.

• Never snorkel alone, and make sure others know where you are going.

• If you are not a certified diver, do not dive. If you were previously certified, but have not been diving in a while, update your training and dive only with a certified instructor.

• Check your equipment fit and function with your snorkeling or diving partner. Use a properly fitted mask and fins. A mask is the correct size if you can keep it tight against your face by inhaling, without the strap around your head. If you cannot get good suction, try another mask. Position straps and snorkel comfortably—not too tight. Remove hair from all edges of your mask to prevent leakage.

• If you are inexperienced, enter the water slowly and practice in the shallows first. Sit in shallow water and put your face in the water to see if your mask fits properly and you are breathing effortlessly. Taking the first few breaths of air through a snorkel while your head is underwater is a little unnerving, but you will soon be breathing with ease if you give yourself time to get used to this new experience.

• Once relaxed, head into deeper water, but do not go beyond your ability level or that of your partner. Snorkel or diving equipment does not make you a better swimmer.

Reef-friendly Behavior

• **Look, but do not touch**. Respect the extremely fragile coral reef environment and preserve this underwater garden for the creatures that live there.

• Avoid touching live coral: a simple touch can kill the coral polyps (see glossary). Instead use algae-covered rubble or rocks to steady yourself, if necessary.

• Tread lightly. Be respectful of the reef and all its inhabitants. The reef is a living thing, and what look like rocks or plants are probably thousands of tiny animals trying to build a place to live. Enter, exit, and rest only on sandy bottoms or bare rock, NOT CORAL. Coral is very fragile. It may take many years to regrow one broken branch of coral.

• Stay far enough above the bottom to avoid making fin contact with coral or kicking up sediment. The sediment smothers delicate invertebrates such as corals and sponges.

• If you move coral rubble looking for invertebrates, replace the rubble EXACTLY as you found it, slowly and carefully, being careful not to squish the animals hiding underneath. The animals live there because they are not able to live in sunlight, currents, or sediment.

• Approach fish slowly and respectfully.

• Do not feed the fish. Artificial foods can be unhealthy and may change the fishes' natural behavior. Larger fish may be attracted to feeding areas and are often aggressive toward those fish that are adapted to living there.

• Leave live animals and plants in the water. Most species need very limited, specific conditions to survive; for example, they may be able to eat only polyps from one type of coral. Others, such as flatworms, nudibranchs, and anemones, are too delicate for transport.

• If collecting shells washed up on the beach, take only small quantities. NEVER collect living shells from their marine habitat.

• Local legends say that terrible things will happen to those who take shells from the island of Ofu, along the area of the beach known as Toaga (now part of the National Park of American Sāmoa). Purchase of shells is not recommended. They may have been gathered alive, using highly destructive techniques.

• Sometimes shell collectors take too many animals of one species. This upsets the delicate balance of the coral reef ecosystem. Overcollection of species can also result in insufficient food for other species, leading to their extinction. In addition, extirpation or extinction may occur when a sufficient number of mating partners can no longer be found.

• Leave seemingly empty shells in the water. Take home only empty shells washed up on the beach. Look carefully; there might be a hermit crab deep within. If you expel the crab, you are probably sentencing the now unprotected animal to death.

• Do not remove or move attached animals or plants to photograph them; it will subject them to stress or kill them.

• Rocks or coral rubble thrown into the water can break or kill delicate coral branches, or perhaps land heavily on and injure or kill an animal such as a sea cucumber.

• If diving, do not stay long in caves or in any one location of a cave. Large amounts of expelled air become trapped, creating ceiling pockets and killing the animals that live there.

• Anchor boats only in sandy-bottomed areas; anchors can cause irreparable damage to corals.

Scientific Classification

Taxonomy (also known as "systematics") is the branch of science that identifies, names, and studies the relationship between organisms. Each animal or plant is given a two-part scientific name; the first is its genus and the second is its species name. The genus and species name are always italicized. All organisms are placed into hierarchical categories based on their relationship to other similar organisms.

The same animal can be known by different common names in different areas, but its scientific name helps avoid confusion. For example, the following common names— fairy tern, white tern, **gogosina** (Sāmoan)—all refer to the same species of bird, *Gygis alba*. Two scientists from different areas of the world, for instance American Sāmoa and Australia, can be assured they are talking about the same organism if they use its scientific name.

Gogosina has the following taxonomic classification:

Kingdom:	Animalia (includes all animals)
Phylum:	Chordata (includes all animals with a notochord or backbone)
Class:	Aves (includes all birds)
Order:	Charadriiformes (includes all shorebirds and waders)
Family:	Sternidae (includes all terns and noddies)
Genus:	*Gygis*
Species:	*alba*

As taxonomists learn more and more about organisms, occasionally a species will be given a new scientific name or placed in a different taxonomic group.

THE CORAL REEF

CORAL REEF ECOLOGY

Coral reefs (**a'au**) are wonders of the underwater world. They contain some of the most colorful and varied forms of life on earth. Like rain forests, they contain a multitude of organisms, which collectively constitute the reefs' biological diversity or biodiversity (**tamaoaiga o meaola ese'ese**). In land communities, plants (**la'au**) outnumber animals (**manu fetolofi**). But even though plants are an important component of reefs, only a few of the organisms growing on a coral reef are true plants (see below). It is a confusing, watery world in which the coral animals often look like plants, creating a coral fairyland with a diversity of animal forms and shapes that rivals the lush plant growth of a tropical jungle. In fact, coral reefs have been referred to as underwater rain forests.

When most people visit reefs, they are captivated by the variety of fishes (**i'a**) found there. Nevertheless, the reef itself is alive with tiny coral polyps that build the reef's coral structures. Corals (**amu**) are actually colonies of tiny animals living together on the reef. The individual animals, some no bigger than the head of a pin, that build the reef are called "coral polyps." Each polyp, typically less than 1 cm (0.5 in.) in diameter, has a tube-shaped body and a mouth surrounded by tentacles. These animals belong to the phylum Cnidaria (formerly Coelenterata). Special cells in the coral polyp absorb calcium, carbon, and oxygen from the water, combining the ingredients into calcium carbonate (limestone). The tiny coral polyps then secrete this limestone in a "cementlike" formation, creating a safe house around each individual. Thousands of coral polyps living together in their limestone homes form the structure we call coral. Each new polyp adds this animal cement to the family home, and the coral colony grows. A reef ecosystem comprises many different species of corals.

Within the body of the coral polyp live small, single-celled algae known as "zooxanthellae." These tiny plantlike cells photosynthesize in their host coral's body. Just like terrestrial plants, they make food using sunlight and nutrients. Photosynthesis uses sunlight to convert carbon dioxide and water into food and oxygen. The coral polyp benefits by sharing the food made by the zooxanthellae. The polyp receives many of the nutrients it needs for basic survival from its algal partner. In return, the zooxanthellae receive a safe place to live within the limestone exoskeleton surrounding the coral polyp as well as protection provided by the polyp's tentacles. Zooxanthellae may also derive some inorganic nutrients they need to photosynthesize from the waste products of the coral polyp. The zooxanthellae are the reason for the variety of colors we observe in different species of corals.

Coral polyps also feed by exposing their tiny stinging tentacles called "nematocysts." The nematocysts capture food particles carried by water moving across the reef. During the day, the polyps of most coral species retreat inside their limestone homes. As night falls, the polyps of many species emerge to feed.

Coral polyps reproduce in several ways: fragmentation and budding (two types of asexual reproduction) and spawning (sexual reproduction). Fragmentation occurs when a piece of coral, broken off the parent colony, lands in a suitable location and continues to grow. Budding is the process whereby a coral polyp produces a genetically identical offspring off its own body. During certain times of the year, when ocean and moon conditions are just right, many species of corals reproduce in a mass event known as spawning (**feusuaina o amu**). The spawning corals release massive numbers of sperm and eggs into the water, producing an underwater "snowstorm." The release of eggs and sperm (collectively known as "gametes") by one coral colony will trigger spawning in neighboring corals of the same species. The spawning event can thus take place synchronously across miles and miles of reef, but different species of corals may spawn on different nights or at different times of the night. The eggs are subsequently fertilized by sperm that are propelled along by currents moving across the reef. Once fertilized, the eggs develop into planula, which is the free-swimming (planktonic) larval stage of coral. The floating planula may travel great distances, or perhaps only to an adjacent reef. If at last it finds favorable conditions, the planula attaches itself to a vacant patch of reef, matures into a polyp, and begins a new coral colony.

How does a coral grow? When a polyp dies, a new one may grow on top of the old one. A coral colony is therefore composed of a thin layer of living polyps growing on a huge number of skeletons of dead ones. Under the live colonies is the reef structure itself, composed of dead colonies cemented together by coralline algae. New colonies and coral structures are constantly being built on top of the dead skeletons of older colonies. The rate of coral growth, nevertheless, is relatively very slow. Branching corals such as staghorn and elkhorn grow only about three inches a year. Massive boulder corals grow even more slowly: a basketball-sized coral may be fifty years old or more.

These structures of the reef serve as underwater housing for a dazzling array of marine life. The maze of openings, grooves, crevices, and hiding places formed by corals provides a safe nursery for young fish and a sanctuary for lobsters, crabs, shellfishes, and a plethora of other marine invertebrates. Intermixed with the hard, stony corals are algae and sponges. Occasional anemones, soft corals, and tube worms flutter in the currents. Extraordinarily patterned and colored fish meander here and there among the corals, creating pulsating rhythms of color and a profusion of fish faces. Microscopic organisms, macroinvertebrates, and a host of other creatures find shelter in crevices and under ledges. Larger predators such as sharks, barracudas, and groupers prowl the reefs in search of food. Moray eels, when not hiding under a coral head waiting to ambush a meal, may be seen undulating their bodies across the reef. The sharp beaks of parrotfishes pulverize coral mouthfuls into fine-grained particles, creating and circulating a major component of beaches: coral sand.

Corals come in countless shapes and provide both a beautiful and functional habitat for coral reef fish and other marine creatures. Most of the plants and animals live and depend on the reef for survival. The reef provides food, shelter, and a safe nursery area for their young. The hundreds of species of corals found in Sāmoa decorate the ocean floor in an infinite variety of colors and intricate patterns (**ituaiga amu ese'ese**). These include the stalks and columns of pillar corals, the fragile tree and branch shapes of stone-hard antler corals, huge coral boulders that look like cauliflower and giant brains, and the delicate, blossom-like forms of lettuce and leaf corals.

Even dead coral heads and pieces of coral rubble provide important habitat to a unique assemblage of species that are sciaphilic (light-avoiding). These species encrust the dark undersides of everything, growing in a profusion of forms and colors. If you accidentally turn over a piece of coral rubble, make sure you gently replace it exactly as you found it; it is most likely some animal's home. The species found underneath cannot live or grow in direct sunlight, and many other reef residents, such as gastropods, nudibranchs, and brittle stars, depend on these hidden species as a source of food.

In summary, the coral reef is a monumental formation constructed over thousands of years—so complex, interrelated, and codependent that one change can negatively impact the health of the entire coral ecosystem.

Because coral communities provide important habitat for so many creatures, deterioration in their health may seriously degrade associated fish and marine fauna. As in other ecosystems, every creature living in and around the reef depends on every other for survival. When one species is lost, its absence can negatively affect all of the other residents. For example, what would happen if overfishing resulted in the loss of many of our herbivorous fishes? When reef grazers such as herbivorous fish decrease in number, the algae are free to grow and spread over large areas, killing or weakening the health of the coral and making it susceptible to disease and other disturbances. This one event could eventually devastate all the residents of that reef.

TYPES OF CORAL REEFS

The coral reefs of the archipelago can best be categorized as fringing reefs, barrier reefs, and coral atolls.

Fringing Reefs

Fringing reefs are in the earliest stage of reef development, while barrier reefs and atolls are in the mature stages. A fringing reef generally occurs close to shore and does not yet contain a well-developed lagoon. Later, as the island itself erodes and sinks, and as the coral grows seaward, a lagoon begins to develop close to shore.

Barrier Reefs

At a later stage in fringing reef development, as the island erodes and subsides into the ocean floor, the reef is now farther from shore and protects an often expansive lagoon. Many of the reefs surrounding the western islands of Sāmoa are barrier reefs. Australia's Great Barrier Reef is the best-known example of an extensive barrier reef system.

Atolls

An atoll develops when a volcanic island surrounded by a fringing reef begins to sink back into the sea floor. While the island is submerging, the coral continues to grow upward. Eventually, the island is completely submerged and only a coral ring remains near the surface.

A coral reef may fit neatly into one of the above categories, or be in a state of maturation somewhere in between. Rose Atoll, an extremely important seabird and sea turtle nesting area, is a small, square atoll.

REEF PROFILE AND HABITAT TYPES

All reefs are different from one another. Some are young; others are old. Further, each reef is influenced by various factors: the currents that move through the lagoon, its exposure to wave action, the amount of runoff from land, and the amount of sediments or other pollutants in the runoff.

Most of the reefs of American Sāmoa are fringing reefs that lie close to shore (< 200 m/219 yd.), while the western islands of Sāmoa are mostly surrounded by barrier reefs. The coral reef itself can be divided into several habitat classes that differ in location relative to the shoreline and exposure to the sea.

Reef (or Rubble) Flat

The reef flat is situated somewhere between the shoreline (if there is an island present) and outer edge of the reef. It is usually exposed during very low tides; coral growth is usually limited to those areas that remain covered by water during even the lowest tide. Its surface can be pavement smooth, covered with coral rubble, or riddled with deeper tide pools that are rimmed with coral. Typical depth is 0–1 m (0–3.3 ft.).

Reef Crest

The reef crest is defined as the seaward edge of the reef flat, where the reef edge plummets into deeper water. It is here that most incoming waves are breaking. Typical depth is 0–3 m (0–10 ft.).

Reef Front

The reef front descends from the reef crest to the reef base, or submarine terrace, at a slope of 45–90 degrees; there is frequently a spur and groove zone (visible from the air), in which the reef is dissected by deep channels. Typical depth is 10–30 m (33–98 ft.).

Lagoon

A lagoon is located between the reef flat and the shore-line. Lagoon habitat is uncommon in American Sāmoa; conversely, the reefs of the western islands of Sāmoa are typically barrier reefs, and the reef flat itself is separated from the shoreline by natural lagoons up to 2 km (1.2 mi.) wide in some locations. Typical depth is 1–3 m (3.3–10 ft.). The area between a fringing reef and the shore is the reef flat, while the area between a barrier reef and the shore is a lagoon.

Pinnacles

These occur only at Rose Atoll; the lagoon floor has numerous flat-topped, steep-sided coral pinnacles that rise from the depth of the lagoon to the surface. At Rose Atoll, the reef flat encircles a lagoon floor (approximately 15 m/49.2 ft. deep) from which tall coral pinnacles rise to the ocean's surface. This lagoon is open to the sea through a deep channel in the reef. While Swains Island (an atoll) has a similar central lagoon, the lagoon is not open to the ocean, and its water is brackish.

THE CORAL REEFS OF SĀMOA

The coral reefs of the Sāmoan Archipelago are biologically diverse. Over 200 species of corals have been reported in the archipelago. More than half of all the coral species known in the entire Indo-Pacific region, which stretches from East Africa to the islands of Polynesia, are found here. In addition, there are approximately 915 nearshore fish species in the waters of the archipelago, compared with only 460 nearshore fish species in Hawaiʻi.

The coral reefs of individual islands in the archipelago differ in the amount of human impact they have been subjected to over the years. The reefs and other habitats found on the islands of Tutuila and ʻUpolu are the most degraded by human impacts. For example, rapid population growth and industrial development in Pago Pago Harbor on Tutuila are placing increasing stress on the reefs. For the most part, the remaining reefs inside Pago Pago Harbor are now badly degraded. While there has been some improvement in water quality during the past decade, without a continued reduction in pollution and sedimentation, the harbor may not recover. Conversely, for the most part, reefs that are remote and removed from many human impacts are still in good condition.

In addition to direct human influences, Sāmoa's coral reefs have been subjected to several other devastating impacts over the last two decades. During the late 1970s, a significant outbreak of the coral-eating crown-of-thorns starfish (*Acanthaster planci*, **alamea**) consumed most of the coral surrounding Tutuila and the western islands of Sāmoa (see Marine Invertebrates chapter). Interestingly, the reefs of Manu'a, Rose, and Swains were spared this particular crown-of-thorns infestation. Just as the impacted reefs were beginning to recover, the following three catastrophic tropical hurricanes occurred: Tusi in January 1987, Ofa in February 1990, and Val in December 1991. Each hurricane affected individual islands in the archipelago differently. The last two hurricanes were particularly devastating to the ecosystems of Tutuila, including the reef. Val was particularly damaging to the island of Tutuila. Ofa was the worst storm to hit the western islands of Sāmoa in 169 years, with windspeeds of 180 kph (112 mph) in Apia, and thought to be much stronger in Falealupo, located on the western end of Savai'i. The intense wave activity of the storms overturned much of the coral near shore and severely damaged corals to a depth of up to 10 m (30 ft.). More recently, in 1994, a mass "coral bleaching" event affected most of the coral in the archipelago (see inset, pp. 22–23).

For a detailed description of the coral reefs of American Sāmoa, see Birkeland et al. (1997), Craig (1996), and Green

(1996); for the western islands of Sāmoa see Morton (1993), and Zann (1991).

෴෴෴෴෴෴෴෴෴෴෴෴෴෴෴෴

Coral Bleaching

"Coral bleaching" is the term used to describe the death of corals (amu mamate) from the loss of their photosynthetic zooxanthellae. Coral bleaching occurs when corals undergo stress from unnaturally warm waters, pollution, increased solar ultraviolet radiation, or other factors. During coral bleaching the corals expel their algal tenants, the zooxanthellae. Without the color-giving algae, corals appear pale and bleached, and they no longer have the benefit of the meals provided by the algae.

Zooxanthellae are believed to abandon their coral hosts when there is a sustained, though slight, rise in sea temperature. Scientists speculate that the rise in sea temperature may be a result of natural cyclic changes in the ocean's currents (e.g., El Niño) or perhaps global warming. A rise in temperature of only two degrees sustained over several days' time can result in bleaching. Bleached corals are not dead initially, but they are unable to grow or reproduce. The corals can, however, be repopulated by zooxanthellae, but they will die if these food-making algae do not return.

How quickly corals can recover from the bleaching event depends upon how much stress the corals experienced and for how long. If the corals are living in clean water, they can rebound. However, if they are subjected to additional stress from pollution, sedimentation, or other factors, they may die while in their bleached state.

Many scientists now connect coral bleaching with global warming, but no one knows for sure.

Over the long term, the cycle of growth and destruction of the reef is a natural cycle observed in tropical marine ecosystems. On the other hand, there is nothing natural about global warming.

This large-scale type of coral bleaching should not be confused with the very destructive and illegal method of fishing in which laundry bleach is poured into the water to kill the fish (**o le fa'aaogaina o vaila'au oona e faaleagaina ai i'a ma amu**). This destructive method kills not only the fish, but also the coral and all other marine creatures living in the area, thus rendering the reef unsuitable for recolonization by fish or any other marine animals.

At present, the reefs and associated marine life in the archipelago are in a significant stage of recovery, and coral colonies are regrowing quickly at most locations around Sāmoa. Nevertheless, the reefs remain vulnerable to threats from human activity and exploitation.

Storms and other seemingly destructive events are natural occurrences in a tropical ecosystem. Occasional natural disturbances may in fact play a role in maintaining the breathtaking diversity we observe on coral reefs because new species may colonize disturbed areas.

Coral Diseases

Coral reefs worldwide are showing signs of stress and ill health. In the last few years, scientists have documented an increasing number of coral diseases, some of which have everyone baffled.

Black-band disease is slow acting and caused by a bacteria that produces a mat of black filaments. These filaments appear to kill the coral by excreting some type of poison. Black-band disease affects many different types of corals and has been observed in several tropical reef locations around the world. White-band diseases kill and bleach the coral tissue. While one white-band disease type is slow acting, the other type progresses across a coral head at the rate of 9 cm a day! This disease is different from the coral bleaching that occurs when corals undergo stress (see inset on coral bleaching), and its cause is unknown. Yellow-blotch and white-plague are additional new diseases afflicting corals.

Still more mysterious and catastrophic is rapid-wasting syndrome. It is unlike other coral diseases, which kill the coral polyps but leave the coral skeleton intact. Instead, the coral skeleton itself crumbles and disintegrates into sand grains when touched, much as a cube of sugar or a sandcastle would. Rapid-wasting syndrome is so-named because it can spread several inches across a coral head in a single day. The cause of this new, disastrous disease is unknown.

THE VALUE OF REEFS

Corals are not only beautiful; they also provide free services that are vital to the protection and economic welfare of humans.

• Way of life: Reefs play a significant role in Sāmoan life by providing over 50% of the local fish harvest. They are also an important connection between traditional culture and the marine environment, and many important legends center on the marine environment or its inhabitants. According to the Sāmoan legend "Turtle and Shark" (which has many variations), a blind woman and her granddaughter were banished from their village on the island of Savai'i during a time of famine. Ending up on Tutuila, in the village of Vaitogi, the two leapt off the cliffs into the sea and turned into a turtle and a shark. The family of the two, guilt-ridden, went to the shore and called their names. When a turtle and a shark appeared, the family then knew the grandmother and grand-daughter were safe. In this village today, Sāmoan children sing to call to the turtle and shark, which often appear.

• Storm protection: The reefs are a first defense against the battering of tropical storms, helping to protect shorelines and villages from wave erosion. The corals themselves form a natural and self-repairing breakwater that shelters the islands against the violent force of the sea. In addition, the reef's porous structure absorbs and dissipates the energy of oncoming waves.

• Sand factory: Corals and calcareous algae (e.g., *Halimeda* spp.) are a major source of sand. Fishes, especial-ly parrotfishes, browsing on these organisms contribute tons of sand per acre every year. Some of this sand replenishes beaches, while the rest becomes part of the reef floor. The sand also builds the outer edges of deeper slopes offshore.

• Tourism: The Sāmoa Islands' beautiful reef ecosys-tems attract sightseers, snorkelers, and divers. The reefs also serve as an important nursery area for many of the fish that attract recreational fishermen.

• Scientific research: It is frequently said that we know more about other planets in the solar system than we do about our ocean environments. Reefs serve as a living labo-ratory for scientists, who learn about the unique creatures of the marine environment. Many chemical compounds found on reefs, and in reef animals, are being studied to see how they might possibly be used in medicine or to improve human health.

THREATS TO CORAL REEFS

As we have learned, reef-building corals are delicate animals with very exact requirements. In general they can-not withstand long exposure to temperatures cooler than 20 degrees C (68 degrees F) or warmer than 32 degrees C (85 degrees F), do poorly on soft or shifting bottom sediments or in constantly turbid water with dim light, and cannot tolerate changes in salinity or water quality.

Overfishing

Overfishing (**soona fagotaina**) is a major cause of coral reef destruction around the world. Overfishing results in fewer fish grazing on the algal mats growing on the reef. Without enough fish, the algae grow quickly and outcompete coral for space and sunlight. This results in the corals' being smothered to death. In Sāmoa, new and sometimes destructive fishing methods, such as nighttime spearfishing using scuba gear, are removing too many fish from the reef (see Parrotfishes).

Deforestation and Sedimentation

Rainwater flows down from the mountains and uplands, first into streams, then into the ocean and onto the reefs. During large storms, landslides often occur, transporting materials quickly downslope into swift-running streams that flow into the sea. If the land has been disturbed or cleared of its natural vegetation, the process is exacerbated and the water carries even larger amounts of mud and other debris down the stream as well. The sediment then settles onto the corals, blocking the light and preventing the algae in the polyps from photosynthesizing, thereby smothering them (**faaleagaina**). In this way, even though the reef may be located a great distance from the deforested land, the sediment created by clearing the land for agriculture, houses, or roads far away can destroy it. In Sāmoa, more and more land is being cleared. When it rains, sediment that blankets the coral and suffocates the reef is very visible. To make matters worse, muddy sediment may not be the only thing carried to the sea by streams. Runoff in Sāmoa is also a source of "nonpoint source pollution." Nonpoint source pollution (**vaileaga**) includes anything unnatural that is washed into streams or storm drains. Such runoff can contain pollutants and chemicals from roads, automobiles, plantations, pig- geries, village households, and industries. Nutrient enrich- ment from human sewage disposal and domestic animal waste contaminates the water by raising bacterial concentra- tions in the animals and fish living there. Industrial, house- hold, and marina boat contamination of Pago Pago Harbor has resulted in inedible and toxic fish.

It seems that no area of Sāmoa is immune from this problem, even uninhabited coastlines. Feral pigs have moved into the forest and are uprooting and killing rain for- est vegetation in many remote areas. This is resulting in the creation of very muddy streams from even the highest, most remote mountaintops (see Mammals chapter).

Careless Visitors

A serious threat to the reef is damage caused by careless or uninformed visitors. Many people come to the reef to sightsee, snorkel, and dive, and the individual damage from boat anchors, illegal specimen collecting, and other visitor activities—when multiplied by many visitors—can be more impact than the fragile reef can withstand. Although a healthy coral might survive such mistreatment, in places where reef health is already compromised, the entire coral colony could die.

Corals protect themselves from too much sun by pro- ducing mucus. We remove this protective covering when we walk on or touch coral. When near the reef, walk on sandy areas only and do not touch coral. A common misconception is that breaking off a piece of coral does no more damage than pruning a tree or shrub. In actuality, regrowth of broken coral takes a very long time, and seemingly insignificant damage such as abrasions, cuts, scrapes, scratches, and breakage invites invasion of algae and other organisms and diseases that can spread rapidly and kill the entire colony. The living tissue of stony corals is only a thin outer skin consisting of thousands of tiny polyps that cover the interior skeleton. A small "wound" in this covering is similar to that in human skin: it can become infected by other organisms. Some of these grow and multiply at astronomical rates, attacking the coral polyps and stripping the coral of its living tissue until only a bleached, white skeleton remains.

One coral disease is caused by blue-green algae. In laboratory experiments, healthy, undamaged coral specimens repeatedly rejected the blue-green algae. Nevertheless, when the coral was cut with a knife and algae placed in the wound, the algae grew and spread until it killed the entire coral colony.

Do not sit or stand on coral. This common practice of fishermen and snorkelers causes breakage of branching corals and injury to the living tissue of coral heads. If you need a rest, use a float or life vest or stand on a sandy bottom only. Many divers conserve air and steady themselves by holding on to the bottom and pulling themselves from place to place. If they grab living coral, they can injure the delicate tissue. Every time they stand up, they may damage nearby corals and any live coral that is under their flippers. Try to stay "horizontal" while exploring the reef. Touch only dead coral or bottom rocks. If too many people use the reef, the damage is compounded, and in some places, we are "loving our reefs to death."

Coral can also be dangerous to people. Because corals are cnidarians, with their characteristic stinging nematocysts, people should avoid touching them. They easily scratch and abrade, embedding the living coral animal in our skin, which often results in slow healing and infection of the wound if it is not cleaned properly.

Swimming and Boating

Even careless swimming, snorkeling, and boating can threaten the health of a coral reef ecosystem. Touching, standing on, or moving a coral boulder to peek underneath can kill the fragile corals and marine organisms hiding below. Boat anchors are capable of great destruction. Anchors should never be dropped directly on a reef, but rather only in sandy patches. Large areas of coral can be destroyed, and boats damaged, because of careless operation in shallow water. Navigation through shallow coral formations should not be attempted.

Coral and Shell Collecting

Do not collect living corals or shells (**aua le ave'esina figota mole faasao**). Living coral is much more valuable than dead souvenirs. If snorkelers and divers take home only memories or photographs of the reef, it will remain as they remembered it, to be enjoyed on a return visit. Otherwise, the population may die out. In the words of conservationist Gerald Durrell, "hundreds of thousands of sea creatures are killed each year so that their shells may gather dust on suburban mantelpieces."

Like many other shellfish, all species of giant clams need a certain number of individuals on a reef in order to reproduce. If the number of clams falls below this critical number, certain essential environmental conditions do not trigger spawning, and, consequently, there may be no new clams to take up residence.

Trash

Unfortunately, the pristine nature of many reefs has been diminished by thoughtless litterbugs. This is true in Sāmoa, where one cannot visit a beach or reef without seeing some type of trash. When on shore, please keep waste out of streams and illegal coastal dumping areas. If boating, return trash to shore and dispose of it properly.

Using Dynamite and Poison to Capture Fish

Using dynamite or poison to capture fish kills more than the fish. It kills the fishes' coral home, it kills their food, and it kills their young. It is like cutting down a coconut tree to get a coconut. Further, if a coral reef is damaged, wave action on the shoreline may increase and erode shoreline habitats.

In American Sāmoa it is unlawful to possess or use explosives, poisonous or intoxicating substances, including laundry bleach, quinaldine, pesticides, traditional fish poisons such as Barringtonia (*Barringtonia asiatica*, **futu**), or electrical devices to catch fish or shellfish.

In a coral reef community, the forces of both growth and destruction are constantly at work. Under healthy, natural conditions, the constructive forces lead, and the reefs continue to grow. Nevertheless, if human damage tips the balance in favor of the forces of destruction, the reefs will die.

CORAL REEF AND NEARSHORE FISHES

INTRODUCTION

It is hard to imagine a coral reef without its fish inhabitants. They exhibit a diverse variety of shapes, sizes, colors, and behaviors as they busily explore every nook and cranny of the reef. The show is endlessly fascinating for anyone who enters their world. The coral reef and its fish fauna depend upon each other for survival. The reef provides food and shelter to many organisms, and in turn, reef organisms fill important ecological roles on the reef.

Approximately 1,000 species of fishes are known from the archipelago. Of these, 915 or so are considered shallow-water or reef-inhabiting species (nearshore). The rest are deep-water bottom and pelagic fish species. Pelagic fish live in the open ocean rather than adjacent to the land. While the diversity of Sāmoan reef fish fauna is spectacular, it is less than that of the Great Barrier Reef, home to approximately 1,500 species. Even more extraordinary is the area comprising the entire Indo-Pacific region (of which the Sāmoan Archipelago is a part), which contains over 4,000 species of fishes.

The various reefs encircling the archipelago have many of the same fish species. Nevertheless, variations in each island's size, shape, and available habitats result in minor differences in species diversity in a particular location. For example, the island of 'Upolu has a greater amount of fresh-water runoff and more extensive mangrove estuaries. The western islands of Sāmoa also have barrier reefs and wider reef flats in general. At the other end of the spectrum, Rose Atoll has neither exposed volcanic substrate nor any fresh-water runoff; it is also quite isolated and remote.

The illustrated fish families that follow are in taxonomic order, from most primitive to most advanced. Fish nomenclature follows Wass (1984). Some Sāmoan fish names stand by themselves, i.e., their only meaning is that describing a particular fish or family of fishes. A few names refer to a particular characteristic of the fish, e.g., large lips. Maximum length from snout to tail-fin edge is provided after the scientific name in parentheses (in centimeters) and follows Randall et al. (1998).

Fishing regulations exist in both Sāmoas. These fishing regulations for specific families are included in this book, but regulations change, so please consult the local fisheries agency for current fish size limits, gear restrictions, and fishing licenses.

NATURAL HISTORY OF CORAL REEF FISHES

Most coral reef fish are small (< 30 cm/12 in. long, e.g., wrasses, gobies). Most are also diurnal, i.e., typically active during the day. Some are nocturnal (active at night). And some are cryptic (concealed by their coloration) species that frequently go unnoticed.

Coral reef fish live out their lives in a highly structured and elaborate, yet immovable coral environment. Some hide in the coral (e.g., moray eels), while others are so perfectly camouflaged that you might never notice them (e.g., stonefishes). Many of the nocturnal species that hide in the coral during the day leave the safety of the coral and become much more active at night. Some species are large and conspicuous (e.g., humphead wrasse, *Cheilinus undulatus*), while others are tiny (e.g., beaked leatherjacket, *Oxymonocanthus longirostris*). The large, conspicuous species are most vulnerable to overfishing (see inset in Parrotfishes section).

Reef fish also exhibit an array of lifestyles. Some species always occur in groups or schools, others are typically observed in pairs, and still others are solitary. Many species are territorial. These fish steadfastly guard an area encompassing something of importance, such as food, shelter, a mate, or a nesting site. A few species are monogamous, but the majority are polygamous.

Reef fish exhibit a variety of reproductive patterns. Most species have the typical male and female sexes. Some species have separate sexes (male or female), while many other species undergo a sex change as part of their development. This latter pattern is called sequential hermaphroditism. Those that begin life as females and change into males are known as protogynous hermaphrodites (e.g., parrotfishes, wrasses); those that begin as males and change into females are proteandrous hermaphrodites (e.g., moray eels, anemonefishes).

The birth of fully developed offspring is rare among the typical bony fishes; it occurs only in a few species (e.g., some sharks, whose skeletons are made of cartilage, not bone). The vast majority of fish species lay eggs. The tiny eggs (usually 1 mm) take about a week to hatch into larvae that, initially, rarely resemble their fish parents.

How do these larval fish get to the reef or colonize new, distant coral reefs? After hatching, most species spend the first few days or months as larvae. During this stage, they drift with the currents and feed on phytoplankton (microscopic marine plants). As larvae, the fish may travel long distances before encountering another reef, since coral reefs are some of the most patchy and isolated habitats on earth. Many larvae never make it to another reef, but die or get eaten as part of the floating plankton. Those that do make it safely to another reef undergo a change of shape and color when they settle. After settlement, most species spend the rest of their lives on the reef. Some reef species settle into other habitats first (e.g., mangrove areas), moving out to the reefs as they grow. As a consequence of their pelagic larval stage, the juvenile fishes on a reef are not the offspring of the adults present at the same reef.

There are two classes of fishes: the most abundant, class Osteichthys, comprises the typical bony fishes. Class Chondrichthys comprises the sharks and rays. The biggest difference between the two classes is that Chondrichthys have a skeleton made of cartilage, not bone. They do not even have scales like most typical fishes, but rather a leathery skin much like sandpaper. The rough texture comes from tiny teethlike structures on the surface called "denticles."

CIGUATERA POISONING

Ciguatera, a very unpleasant type of fish poisoning, is rare in the archipelago. The incidence of ciguatera varies with area and time of year, and the poison has been found in carnivorous, herbivorous, and detritus-feeding fish. A fish containing ciguatera toxin in its flesh can cause severe illness when eaten, even if the fish is fresh and well cooked.

The poison has been linked to an algae-like organism (diatom) consumed by herbivorous reef fish. These smaller, algae-eating fish are themselves eaten by larger carnivorous fish. As more and more fish are eaten, the poison concentrates in the flesh and organs of the larger carnivore. This concentration process is known as "bioaccumulation." Ciguatera poisoning is most frequently associated with consumption of snappers (**mū**), groupers (**gatala**), barracudas (**sapatu**), and moray eels (**pusi**). Ciguatera poison should not be confused with tetrodotoxin, a poison found in pufferfishes (**sue**).

Symptoms of ciguatera poisoning appear from one to thirty hours after the fish is eaten. A person with mild poisoning might experience weakness and diarrhea and never suspect ciguatera. More serious symptoms can include tingling sensations in the mouth, lips, and throat, or in the soles of the hands and feet, accentuated by contact with cold; nausea; complete malaise; extreme weakness; joint pain; severe itching; and confusion of the sensations of hot or cold.

Extreme cases include progressive paralysis of body muscles, coma, and respiratory or cardiac failure.

If you have suffered from ciguatera, or suspect a poisoning has occurred, do not eat fish that are possibly ciguateric for at least three months, because a second poisoning event could be even more severe.

SHARKS (CARCHARINIDAE, HEMIGALIEDAE, AND SPHYRNIDAE)

Sensational stories (unfortunate ones if you happen to be a shark) perpetuated by movies and television give us the impression that all sharks are vicious predators who spend their time searching the oceans looking for a human meal. Nothing could be farther from the truth. Automobiles, dogs, insects, and lightning are more dangerous and cause more deaths each year than all the sharks in the ocean. The sharks you are likely to encounter around the reefs of the Sāmoa Islands are typically benign and timid. Nevertheless, while rarely encountered, tiger (*Galeocerdo cuvier*, **naiufi**) and hammerhead (*Sphyrna lewini*, **mata'italiga**) sharks should be avoided.

Sharks feed mostly on fish, crustaceans, and mollusks. They have excellent senses of eyesight and smell and can detect the very faint electric fields emitted by fish on the reef. Shark eyes contain special structures that increase their sensitivity to light and thus enhance their ability to see at night.

Sharks exhibit a variety of reproductive modes and types of embryonic development. Some embryos develop freely in the shark's body, some are attached to a placenta, and others are contained in leathery egg cases deposited in marine vegetation. Such egg cases are known as "mermaids' purses." Most sharks give birth to live young in broods ranging from a few up to 100 individuals. In general, though, most species reproduce slowly, and commercial fishing has dangerously reduced their numbers worldwide. Sharks are caught and killed for senseless reasons, such as unwarranted fear or to provide the main ingredient in shark fin soup. If destined for the soup pot, the shark's fins are often sliced off the still-very-much-alive animal, and the shark is then thrown back into the ocean to die a slow, and probably painful death. It cannot swim anymore, and its blood attracts predators. As scientists continue to collect alarming data on decreasing shark populations, many areas are now restricting commercial fishing, and shark-fishing tournaments are no longer taking place.

Thirteen species are known from the archipelago. The general Sāmoan name for sharks is **malie**, though many have species-specific names.

Blacktip reef shark, *Carcharhinus melanopterus* (180 cm/71 in.)
apeape, **malie alamata** (family Carcharhinidae)

Whitetip reef shark, *Triaenodon obesus* (210 cm/82.7 in.)
malu (family Hemigaleidae)

The whitetip reef shark is one of the most commonly observed sharks on Sāmoan reefs. This species rarely reaches 2 m (6 ft.) in length. It is not considered dangerous, and most individuals will swim off at their first glimpse of humans.

Scalloped hammerhead, *Sphyrna lewini* (420 cm/165 in.) **mataʻitaliga**, "eyes on the ears" (family Sphyrnidae)

GUITARFISHES (RHINOBATIDAE)

The guitarfishes are rays with an elongate body similar to that of a shark. These rays are usually found on soft sand or muddy bottoms, often close to shore. They feed on benthic invertebrates and occasionally fishes. Guitarfishes are considered harmless. Only one species of guitarfish is known from the archipelago, though it is bottom-dwelling and cryptic, and could be easily overlooked. The general Sāmoan name for all rays is **fai**.

Giant guitarfish (also known as shovelnose ray), *Rhynchobatus djiddensis* (305 cm/120 in.) **fai**

STINGRAYS (DASYATIDAE)

The stingrays are characterized by a rounded disc or diamond shape that is usually twice as wide as it is long. Most are observed near muddy or sandy-bottomed habitats, rather than near coral reefs. The tail spine is capable of inflicting an extremely painful wound; some fatalities have occurred. Take care when wading on sandy bottoms, and avoid exploring in bare feet. If stepped on, a stingray can whip its tail forward and plunge the spine into its victim with great rapidity (obtain medical assistance). Stingrays bear live offspring that look like miniature adults.

Two species are known from the archipelago.

Kuhl's stingray, *Dasyatis kuhlii* (70 cm/28 in. wide)
fai tala, fai malie

EAGLE RAYS (MYLIOBATIDAE)

The eagle ray's protruding head, extending forward from its disclike body, distinguishes it from its relatives the electric ray, stingray, and manta ray. The wings are triangular, and the tail is much longer than the disc itself. The eagle ray uses its powerful jaws to crush the hard shells of its molluscan prey.

All rays are harmless if left alone, but eagle rays and stingrays do have venomous spines on their whiplike tails and are capable of inflicting very painful wounds (seek medical attention).

One species of eagle ray is known from the archipelago.

Spotted eagle ray, *Aetobatus narinara* (250 cm/98 in. wide)
fai'pe'a or **fai manu** ("ray bat" or "ray bird")
Usually solitary, this striking polka-dotted ray can be seen "flying" over the coral reef or hovering below the ocean surface just beyond the breaking waves.

MANTA RAYS (MOBULIDAE)

Manta or "devil" rays have occasionally been seen in Sāmoan waters. In spite of their name and enormous size, they are gentle creatures. The largest mantas may reach a width of 6.7 m (22 ft.) and weigh over 1,800 kg (3,968 lbs.). Mantas occur singly or in small groups. They feed by straining tiny zooplankton and bait fish with their huge open mouths as they move through the water.

One species is known from the archipelago.

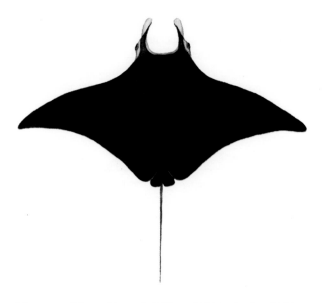

Manta ray, *Manta birostris* (670-cm/264-in. body width)
fai malie

MORAY EELS (MURAENIDAE)

Looking much like a snake, the moray is actually a fish, characterized by a very elongated body. Morays lack pectoral fins, and their dorsal and anal fins join with the caudal fin. Some species have numerous sharp teeth, while others have blunt teeth. Morays have protruding tubular nostrils and modified gill openings that have become holes instead of slits. Some feed primarily on fish, others on crustaceans. All species were once thought to be nocturnal, but only some species forage at night.

While morays can bite humans, their bad reputation is undeserved. Some species open and close their mouths constantly, adding to their somewhat vicious appearance, but these movements are just the morays' way of pumping water across their gills. They will actually take food gingerly and gently from a diver's hand (but feeding morays isn't recommended). Unless it is provoked, a moray is very unlikely to bite. Nevertheless, a moray could mistake a hand placed into a hole for a meal or perceive it as a threat.

In contrast to many other fish families (see Wrasses and Parrotfishes), morays begin life as males, then change into females when they get older (a process known as "proteandry").

At least 46 species are known from the archipelago. Morays are a food fish, but may be ciguatoxic. The general Sāmoan name for moray eels is **pusi**, but names vary with size and color. The smallest morays are called **to'e**, larger ones **maoa'e**, and the largest **atapanoa**. In addition, small brown eels are referred to as **u'aulu**, while pale ones are **apeape**.

Starry moray, *Echidna nebulosa* (70 cm/28 in.)
ai'aiuga

Ringed moray, *Echidna polyzona* (60 cm/24 in.)
The spaces between the dark brown bars are not as wide as the bars themselves; compare with banded moray (*Gymnothorax ruppelliae*), below.

Zebra moray, *Gymnomuraena zebra* (150 cm/59 in.)
to'etapu

Dark-spotted moray, *Gymnothorax fimbriatus* (80 cm/32 in.)
papata pulepule, pusi pulepule

Yellowmargin moray, *Gymnothorax flavimarginatus* (120 cm/47 in.)
tafi laotalo, pusi gatala
Yellowish, but covered with dark brown mottling; compare with giant moray (*Gymnothorax javanicus*), below.

Giant moray, *Gymnothorax javanicus* (220 cm/87 in.)
gatala uli, pusi gatala, pusi maoa'e
Dark spots may develop pale centers in large adults; compare with yellowmargin moray (*Gymnothorax flavimarginatus*), above.

Whitemouth moray, *Gymnothorax meleagris* (100 cm/39 in.)

Banded moray, *Gymnothorax ruppelliae* (80 cm/31 in.)
papata tusitusi
The spaces between the dark bars are about the same width as the bars themselves; compare with ringed moray (*Echidna polyzona*), above.

Undulated moray, *Gymnothorax undulatus* (100 cm/39 in.)
pusi pulepule

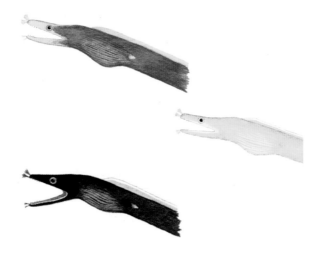

Ribbon eel, *Rhinomuraena quaesit* (130 cm/51 in.)

Sidea picta, common at Rose Atoll.

SNAKE EELS (OPHICHTHIDAE)

Snake eels have a typical eel-shaped body, with tiny eyes located just above the mouth. The snout, often pointed in and downward, is used to burrow into the sand, where the eel spends most of its time. While some species are easily mistaken for sea snakes, snake eels differ from sea snakes in lacking scales and in having tails that are pointed, rather than paddle-shaped (see Reptiles and Amphibians chapter).

Nineteen species are known from the archipelago. There is no general Sāmoan name for snake eels.

Marbled snake eel, *Callechelys marmorata* (57 cm/22 in.)

Culverin snake eel (also known as saddled eel), *Leiuranus semicinctus* (60 cm/24 in.)
gatauli

Harlequin snake eel, *Myrichthys colubrinus* (88 cm/35 in.)
gatamea

Spotted snake eel, *Myrichthys maculosus* (50 cm/20 in.)

CONGER EELS (CONGRIDAE)

Similar to moray eels, conger eels have a round, tubular body and pectoral fins. The most well known species live in colonies in the sand and are known as "garden eels." Their bodies emerge like the stems of tulips when feeding on plankton carried by the currents. Other species favor deep-water habitats.

Spotted garden-eel, *Heteroconger hassi* (40 cm/16 in.)

LIZARDFISHES (SYNODONTIDAE)

The fishes in this family have a some-what reptilian appearance, hence the name. The caudal fin is forked, and most species have irregular spots and blotches on their sides. Most species are found on sandy or muddy bottoms, where they bury themselves with only their eyes showing. While hidden in this way, they are able to lunge upward to quickly seize small fish or crustaceans that swim by.

Five or so species are known from the archipelago. As it is for sandperches (Pinguipedidae), the general Sāmoan name for lizardfishes is **ta'oto**. The species illustrated are frequently observed in coral reef habitats.

Javelinfish, *Synodus jaculum* (20 cm/8 in.)
This species is usually observed in pairs.

Reef lizardfish, *Synodus variegatus* (28.3 cm/11 in.)

ANGLERFISHES (ANTENNARIIDAE)

These rotund, bumpy, highly camouflaged fishes, also known as "frogfishes," have very large, upward-turned mouths. The pectoral fin is limblike, serving to prop the fish into fin position. Their first dorsal spine, called an "illicium," has been greatly modified into a small "fishing pole" tipped with a tempting lure. The anglerfish wiggles this lure in a circular pattern resembling a swimming motion to attract smaller fish prey, which it then swallows whole. Members of this family are also known as "frogfishes."

Seven species are known from the archipelago. They are known by the same general Sāmoan names as those in the family Scorpaenidae: **la'otale** (< 8cm), **nofu** (> 8 cm).

Freckled anglerfish, *Antennarius coccineus* (13 cm/5 in.)

Giant frogfish, *Antennarius commersonii* (26.5 cm/10 in.)

Spotfin frogfish, *Antennarius nummifer* (13 cm/5 in.)

NEEDLEFISHES (BELONIDAE)

The needlefishes, as their name suggests, are long, slender-bodied fishes with extremely elongate, needlelike jaws filled with needlelike teeth. They spend their time just under the surface of the water. Their bluish dorsal and silvery ventral coloration make them difficult to distinguish from above or below, allowing them to blend perfectly into their watery world. It is easy not to notice them if one does not occasionally glance up toward the surface while snorkeling.

Needlefishes (also known as "longtoms" or "houndfish") feed primarily on smaller fishes and are edible themselves. They are voracious predators whose speed and agility enable them to catch prey or escape larger predators. When frightened, a needlefish may even leap out of the water. Human fatalities resulting from impalement by needlefish as they leap out of the water have been reported.

Four species are known from the archipelago. The Sāmoan name for needlefishes varies with size; smaller individuals (< 40 cm) are known as **ise**, larger individuals as **aʻu**.

Flat-tailed longtom, *Platybelone argalus platyura* (45 cm/18 in.)
ise, aʻu
This species is circumtropical.

FLYINGFISHES (EXOCOETIDAE)

Flyingfishes earned their name from their ability to emerge from the ocean and glide above the waves on wing-like pectoral fins for distances up to 300 meters. They are pelagic and prefer deep-water areas, where they feed on plankton. Their eggs have special filaments that stick to floating plant matter until hatching.

Fifteen species are recorded for the area. The general Sāmoan name for flyingfishes is **mālolo**.

Sutton's flyingfish, *Cypselurus suttoni* (30 cm/12 in.)
mālolo

SQUIRRELFISHES (HOLOCENTRIDAE)

The squirrelfishes, also known as "soldierfishes," are found on coral reefs in areas with rocky bottoms. They are spiny, with very large eyes and mostly red coloration. As their large eyes imply, they are nocturnal. During the day they hide under ledges or in caves. If you happen to be looking beneath a coral ledge and see a reddish fish with huge eyes, it is probably a squirrelfish. They are among the most abundant nocturnal fish on the reef.

Most species feed on crustaceans. Some species, such as those from the genus *Myripristis*, are able to produce a variety of clicks, grunts, and growls. If handling them, be careful, as squirrelfishes have a venomous spine on the preopercle (gill flap) that can inflict a painful wound.

Larger species are a food fish on most tropical islands.

Thirty-one species are known from the archipelago. The general Sāmoan name for squirrelfishes is **malau**.

Bigscale soldierfish, *Myripristis berndti* (30 cm/12 in.)
malau ugatele, **malau va'ava'a**

Spotfin squirrelfish, *Neoniphon sammara* (30 cm/12 in.)
malau tui, **malau pe'ape'a**

Crown squirrelfish, *Sargocentron caudimaculatum* (21 cm/8 in.)
malau i'usina, **tāmalau mūmū**, **tameno mūmū**

Sabre squirrelfish, *Sargocentron spiniferum* (45 cm/18 in.)
tāmalau (< 30 cm/12 in.), **mu malau** (> 30 cm/12 in.),
malau toa

TRUMPETFISHES (AULOSTOMIDAE)

Only one species of trumpetfish occurs in the Indo-Pacific. Trumpetfish are elongate and laterally compressed, with an elastic, tubular snout, hence their name. Trumpetfish have a small chin barbel and several short dorsal spines. They are cunning predators, sneaking up on smaller fish and quickly vacuuming them into their mouth. It is reported that the elastic tissue of the snout is capable of stretching and expanding to fifty times its normal volume, an action that sucks the prey into the trumpetfish's mouth.

Stalking predators, such as the trumpetfishes, needle-fishes, and barracudas, have elongate bodies that present a minimal profile (somewhat like a ruler viewed head-on) when they are in position to strike at their prey. Trumpetfish sometimes zoom in on their prey from a vertical position, or orient themselves at an angle that matches that of something such as the nearby branches of an antler coral. Their fins are shaped like the vanes on an arrow, and the attack comes quickly. While there are generally two color phases (brown and yellow), trumpetfish can modify their coloration to blend in with their surroundings or to match the coloration of the very species they are trying to devour.

One species occurs in the archipelago. There are several Sāmoan names for the trumpetfish: **taoto ena** (brown phase), **taoto sama** (yellow phase), **ʻauʻaulauti**, **taotito**. Edibility is poor.

Trumpetfish, *Aulostomus chinensis* (80 cm/31 in.)

FLUTEMOUTHS (FISTULARIIDAE)

The flutemouths (also known as cornetfishes) are similar in shape to the trumpetfishes, except that their bodies are vertically compressed. Flutemouths are more elongate and have very long, tubular snouts with tiny, somewhat lopsided mouths. A single dorsal spine is located directly over the anal fin, and a long filament extends beyond the tail. Like trumpetfishes, flutemouths feed by quickly sucking prey into their mouths. They are usually seen in open, sandy areas and are circumtropical in distribution.

One species is known from the archipelago.

Smooth flutemouth, *Fistularia commersonii* (150 cm/59 in.)
taoto ama, taotao

PIPEFISHES AND SEAHORSES (SYNGNATHIDAE)

Pipefishes belong to the same family as seahorses. They have long, extremely slender bodies, with very delicate, almost invisible fins. Unlike seahorses, they do not hold their bodies vertically. Pipefishes lack pelvic fins and are poor swimmers. Most species are secretive, but those of the genus *Corythoichthys* are often seen in the open. They feed by using their mouths like a pipette, sucking small crustaceans into their mouths.

The most unusual aspect of this family's biology is egg incubation by the male. The female of the species deposits her eggs in a pouch on the ventral surface of the male. The eggs are then carried by the "expectant" male until hatching.

At least 16 species are known from the archipelago. There are no Sāmoan names for pipefishes, probably because of their small size and generally secretive nature.

Although seahorses are not known to occur in Sāmoan waters, there is one unconfirmed record of a spotted seahorse (*Hippocampus kuda*) in the archipelago.

Banded pipefish, *Corythoichthys intestinalis* (16 cm/6 in.)

Ringed pipefish, *Doryrhamphus dactyliophorus* (18 cm/7 in.), IUCN Red List: DD

Bluestripe pipefish, *Doryrhamphus excisus excisus* (7 cm/3 in.)

SCORPIONFISHES (SCORPAENIDAE)

Scorpionfishes get their name from the extremely venomous spines that occur on most species. Members of this family are also known by several other common names: stonefish, toadfish, frogfish, lionfish, firefish, and leaffish. These fishes usually have a single dorsal fin that is strongly notched along the spinous portion. Scorpionfishes are bottom-dwelling predators and are secretive in nature. Some species are extremely well camouflaged, blending perfectly into the substrate, which makes them even more dangerous. When lying motionless on the bottom, they are almost impossible to distinguish from their surroundings.

All members of the family have bizarre appearances. Some scorpionfishes have long, elaborate spikes and tassels, while the stonefishes typically have well-camouflaged, fringed, warty bodies with bulbous eyes. The stonefish's lumpy body resembles the texture of the sponges and rubble upon which it hides. The skinflaps of the leaf scorpionfish are actually modified scales that help to camouflage it, while the weblike pectoral fins of lionfishes (also called firefishes) are sometimes used to push smaller fish into a corner, where they can then be easily captured.

The dorsal, anal, and pelvic spines of the fishes in this family are venomous. The venom is produced by glandular tissue along the spines. Pain from a scorpaenid injury ranges in severity from that of a bee sting to unimaginable agony. If stung, immerse the wound into very hot water to relieve the pain and seek medical attention immediately.

At least 20 species are known from the archipelago. The Sāmoan name varies with size; small scorpionfishes (< 8 cm) are known as **la'otale**, while larger ones (> 8 cm) are known as **nofu** ("something thorny") or **i'atala** ("rough fish").

Red firefish, *Pterois volitans* (38 cm/15 in.)
sausau lele

False stonefish, *Scorpaenopsis diabolus* (30 cm/12 in.)

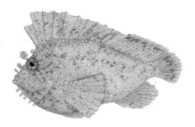

Leaf scorpionfish, *Taenianotus triacanthus* (10 cm/4 in.)

GROUPERS (SERRANIDAE)

Groupers are a well-known food fish on tropical islands. Groupers have robust bodies, large mouths and lips, and lower jaws that extend beyond the upper jaws. They generally have single dorsal fins, which may be deeply notched in the center. Some members of this family are called "rock-cods" and "sea basses." Most groupers are thought to be sequential hermaphrodites, maturing first as females, then changing to males later in life.

Their large size and curious disposition make them very vulnerable to overfishing (see also Parrotfishes). Once the larger male individuals are overfished, only the smaller females are left, thus destroying the fishes' reproductive potential. If left unhunted, some groupers may live for decades. All are carnivorous and feed on smaller fish and crustaceans.

The colorful fairy basslets are a subfamily of the groupers (e.g., *Anthias* sp.). They are typically much smaller than the groupers and inhabit the outer slope of the reef. They gleam in shades of pink, orange, violet, and magenta. Soapfishes, also in the grouper family, produce a skin toxin, grammistin, which renders them unappetizing to predators. Like the toxin of boxfishes, grammistin can kill other fish confined with a soapfish in a small space such as an aquarium.

At least 38 species of Serranids are known from the

archipelago. Although groupers in some areas may be cigua-toxic (e.g., coronation trout), they are an important food fish. It is illegal to fish and keep groupers less than 20 cm (7.9 in.) in the western islands of Sāmoa. The general Sāmoan name for small groupers (< 30 cm/12 in.) is **gatala**; for medium-sized, **ataʻata** ("smile" or "grin"); and for the largest specimens, **vaolo**.

Peacock rockcod, *Cephalopholis argus* (40 cm/16 in.)
gatala uli, loi

Leopard rockcod, *Cephalopholis leopardus* (20 cm/8 in.)
gatala sina, mataʻele

Coral cod, *Cephalopholis miniata* (41 cm/16 in.)
gatala mumu

Flagtail rockcod, *Cephalopholis urodeta* (27 cm/7 in.)
mataʻele

This species is most commonly observed around Swains Island.

Hexagon rockcod, *Epinephelus hexagonatus* (30 cm/12 in.)
gatala aʻau

Dwarf spotted rockcod (also known as the honeycomb grouper), *Epinephelus merra* (27.5 cm/10.8 in.)
gatala aloalo, gatala pulepule

Thinspine rockcod, *Gracilia albomarginata* (40 cm/16 in.)
Within the archipelago, this fish has been observed only at Rose Atoll.

Sixline soapfish, *Grammistes sexlineatus* (27 cm/11 in.)
taili, tusiloa

Waite's splitfin, *Luzonichthys waitei* (7 cm/3 in.)
This species dominates the reefs at Swains Island, but has not been observed anywhere else in the archipelago.

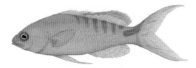

Lori's anthias, *Pseudanthias lori* (12 cm/5 in.)

Amethyst anthias, *Pseudanthias pascalus* (17 cm/7 in.)
segasega moana

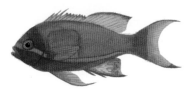

Squarespot anthias, *Pseudanthias pleurotaenia* (20 cm/8 in.)
Females are yellow.

Coronation trout, *Variola louti* (80 cm/31 in.)
papa tuauli (juveniles), **velo** (subadults), **papa** (adults)

TIGER PERCHES (TERAPONIDAE)

Tiger perches, also known as grunters, are perchlike fishes with oblong, compressed bodies and deeply notched dorsal fins. Most species in this family are freshwater fishes. While marine species are not generally observed on reefs, the illustrated species frequents the wave-swept areas of the airport lagoon in American Sāmoa.

A single species is known from the archipelago.

Crescent perch, *Therapon jarbua* (32 cm/12 in.)
avaava

FLAGTAILS (KUHLIIDAE)

These are small, silvery fishes with deeply notched dorsal fins. The caudal fin is conspicuously marked with black, hence the name "flagtail." Juveniles are often found in tide pools or in areas of surge close to shore.

Four species are known from the archipelago.

Banded flagtail, *Kuhlia mugil* (20 cm/8 in.)
safole

CARDINALFISHES (APONGONIDAE)

The cardinalfishes, also called "glassfishes," are usually small and laterally compressed. As the common name

implies, some cardinalfishes are reddish. However, in actuality, most are drab, striped, or silvery. They are usually solitary or occur in small groups or pairs. This family of fishes
is one of the largest, with the majority of species occurring
in the Indo-Pacific region. Cardinalfishes are carnivorous
and feed on zooplankton. They are also nocturnal. It is usually necessary to peek into reef holes and under ledges to
spot a cardinalfish during daylight hours. Once the lights go
out on the reef as night falls, the cardinalfishes emerge from
their hiding places.

The cardinalfishes are unusual in that the males brood
the eggs in their mouths. During spawning, the female
releases an egg mass, which is subsequently fertilized by the
male. He then takes the egg mass into his mouth and does not
feed until the eggs hatch a few days later. Look closely at the
next cardinalfish you encounter to see if you can see the
swollen throat of an egg-brooding male.

At least 30 species are known from the archipelago. The
general Sāmoan name for cardinalfishes is **fō**.

Nine-banded cardinalfish, *Apogon novemfasciatus* (9
cm/3.5 in.)

Five-lined cardinalfish, *Cheilodipterus quinquelineatus* (12
cm/5 in.)
fō tusiloloa

SAND TILEFISHES (MALACANTHIDAE)

The sand tilefishes have elongate bodies. They live on and around sandy or rubble bottoms, often at the base of outer reef drop-offs. If danger approaches, they seek refuge in burrows.

Two species are known from the archipelago. The general Sāmoan names for tilefishes are **mo'o** or **mo'otai**.

Blue blanquillo, *Malacanthus latovittatus* (35 cm/14 in.)
mo'o moana ("envy of the sea" or "gecko of the sea")
This species is often observed in barren sandy areas.

TREVALLIES (CARANGIDAE)

Trevallies, also known as "jacks," are a diverse family. All are streamlined and laterally compressed and have distinctly forked caudal fins. Trevallies are midwater swimmers that frequently occur in large schools. They are commonly seen on reef edges. The offshore species are the fastest, most voracious fishes of the ocean. Their speed, strength, and endurance make them one of the toughest fighters for fishermen.

Twenty-five species are known from the archipelago. Although some species in the genus *Caranx* may be cigua-toxic when large, they are nevertheless an important food fish. It is illegal to fish and keep trevallies less than 25 cm (9.8 in.) in the western islands of Sāmoa. Many species do not have specific Sāmoan names. Instead, each size class is distinguished: **lupo** (< 8cm), **lupotā** (8–20 cm), **malauli** (21–50 cm), **ulua** (51–80 cm), and **sapo'anae** (> 80 cm).

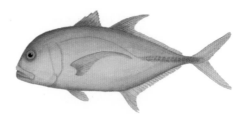

Giant trevally, *Caranx ignobilis* (170 cm/67 in.)
sapoʻanae

Black trevally, *Caranx lugubris* (80 cm/31 in.)
tafauli

Bluefin trevally, *Caranx melampygus* (100 cm/39 in.)
malauli apamoana, atugaloloa

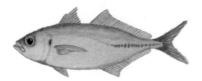

Purse-eyed scad, *Selar crumenophthalmus* (30 cm/12 in.)
nato (< 10 cm), **atule** (10–20 cm), **taupapa** (> 20 cm)

DOLPHINFISHES (CORYPHAENIDAE)

The dolphinfishes should not be confused with the better-known marine mammals of the similar common name (e.g., spinner or bottlenose dolphins), which are somewhat large pelagic species. The species illustrated is characterized by an elongate, compressed body with a long dorsal fin that extends almost its entire length. Adult males develop a bony crest on the forehead.

One species is known from the archipelago. It is recognized worldwide in many restaurants by its Hawaiian name, *mahimahi*; its phonetically similar Sāmoan name is **masimasi**.

Common dolphinfish, *Coryphaena hippurus* (200 cm/79 in.) **masimasi**

SNAPPERS (LUTJANIDAE)

Snappers are generally small to medium fishes with moderately compressed ovate to elongate bodies and single dorsal fins that may be deeply notched. They have sloping, shovel-nosed heads. Snappers can be found in both shallow and deep water (up to 100 m) near reef habitats. Nevertheless, they are wary fish and not frequently encountered by the casual snorkeler or diver. Some may be long-lived, up to 20 years.

Snappers are predators that feed primarily on other fish. Although snappers are an important food fish, they are frequently implicated in ciguatera food poisoning.

At least 25 species are known from the archipelago. While the general Sāmoan name for snappers found in shallow water is **mū**, the name for larger, deep-water species is **palu**.

Small-toothed jobfish, *Aphareus furca* (40 cm/16 in.)
palu aloalo

Green jobfish, *Aprion virescens* (100 cm/39 in.)
asoama, utu

Flame snapper, *Etelis coruscans* (120 cm/47 in.)
palu loa, palu malau, palu atu

Red bass, *Lutjanus bohar* (75 cm/29 in.)
mū, mū aʻa

Yellow-margined seaperch, *Lutjanus fulvus* (40 cm/16 in.)
tamala, tāiva

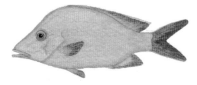

Paddletail, *Lutjanus gibbus* (50 cm/20 in.)
mala'i

Bluestripe seaperch, *Lutjanus kasmira* (35 cm/14 in.)
savane

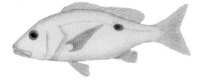

Onespot seaperch, *Lutjanus monostigma* (50 cm/20 in.)
tāiva, **feloitega**

juvenile

Midnight seaperch, *Macolor macularis* (55 cm/22 in.)

juvenile

Black and white seaperch, *Macolor niger* (55 cm/22 in.)
matala'oa

FUSILERS (CAESIONIDAE)

The fusilers are closely related to the snappers and, until recently, were included in that family (Lutjanidae). They are usually observed in schools, and feed on midwater zoo-plankton. They are most commonly observed along the edges of outer reef drop-offs, and often swarm around divers. At night, they return to the reef to sleep, at which time their daytime white bellies often take on a reddish or pink hue. The ventral area also turns red upon death.

The species below is most common around the island of 'Upolu. The general Sāmoan names for fusilers are **atule toto** or **ulisega**.

Red-bellied fusiler, *Caesio cuning* (25 cm/10 in.)
ulisega

SWEETLIPS (HAEMULIDAE)

This family is closely related to the snappers (family Lutjanidae), but their thicker lips, surrounding smaller mouths, impart their gentle, sweet appearance. Sweetlips, also known as "grunts," frequently make "grunting" sounds by grinding their teeth and are one of the noisiest inhabitants of the reef. Most species undergo dramatic color changes as they mature, and the young have striking stripes or spots. Most sweetlips are nocturnal and feed on benthic inverte-brates.

Three species are known from the archipelago. They are an important food fish—not only their lips are sweet! Each species has its own Sāmoan name.

Oriental sweetlips, *Plectorhinchus orientalis* (72 cm/28 in.)
mutumutu or **avaʻava moana**

EMPERORS (LETHRINIDAE)

The emperors are medium-to-large fishes related to the snappers (Lutjanidae). They have thick lips and a single dorsal fin. Most species are observed on sandy patches of the reef. All are carnivorous, and most are nocturnal. Those that feed on other fish may be ciguatoxic.

Thirteen species are known from the archipelago. Each size class has its own Sāmoan name: **mataʻeleʻele** (< 15 cm), **ulamalosi** (15–30 cm), and **filoa** (> 30 cm). It is illegal to fish and keep emperors less than 20 cm (7.9 in.) in the western islands of Sāmoa.

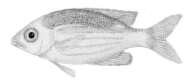

Gold-lined sea bream, *Gnathodentex aurolineatus* (30 cm/12 in.)
mumu, tolai

Thumbprint emperor, *Lethrinus harak* (60 cm/24 in.)
filoa vai

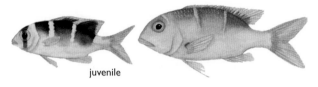

juvenile

Bigeye emperor, *Monotaxis grandoculus* (60 cm/24 in.)
mū matavaivai, **matāmu** (< 15 cm), **matamatāmu** (> 15 cm), **loalia**
The bigeye emperor is capable of altering the brilliancy of its white vertical stripes in the dorsal area, depending on the fish's location and mood. This species is commonly observed during the day.

CORAL BREAMS (NEMIPTERIDAE)
These relatively small, brightly patterned or colored fishes have slender or ovate bodies with single dorsal fins. The breams are usually observed singly or in small groups near sand- or rubble-covered areas of the reef. Breams are diurnal and carnivorous. Although small, they are a tasty food fish.

At least two species are known from the archipelago. The general Sāmoan name for coral breams is **tivao**.

Threelined monacle bream, *Scolopsis trilineatus* (25 cm/10 in.)
tivao

GOATFISHES (MULLIIDAE)
Goatfishes have moderately elongate bodies and two well-spaced dorsal fins. The caudal fin is always forked. Two long, tactile sensory "barbels" located under the chin are

used to probe and locate food items in the substrate. All goatfishes are carnivorous, feeding mostly on invertebrates such as worms, brittlestars, and crustaceans hiding in the sediment and leaving behind clouds of sand as they move along in their search for food. Males also wriggle the whiskerlike barbels during courtship. When not using their barbels, goatfishes can hide them under their gill covers.

Goatfishes are a popular food for humans as well as for other fish. Thirteen species are known from the archipelago, and most individual species have their own name.

Yellowstripe goatfish, *Mulloidichthys flavolineatus* (40 cm/16 in.)
i'asina (< 8 cm), **vete**, **afulu**, **afolu**
This species is most commonly encountered around the island of 'Upolu.

Yellowfin goatfish, *Mulloidichthys vanicolensis* (38 cm/15 in.)
i'asina (< 8 cm), **vete**, **afulu**, **afolu**

Doublebar goatfish, *Parupeneus bifasciatus* (35 cm/14 in.)
matūlau moana

Goldsaddle goatfish, *Parupeneus cyclostomus* (50 cm/20 in.)
moana

Manybar goatfish, *Parupeneus multifasciatus* (30 cm/12 in.)
matūlau, moana

SWEEPERS (PEMPHERIDAE)

The sweepers have moderately deep, angular bodies, with highly tapered tails, oblique mouths set in a frown, and projecting lower jaws. Sweepers are nocturnal, hence their large eyes, and hide in caves or along ledges of surge channels during the day. They emerge at night to feed on zooplankton.

Two species are known from the archipelago. There is no general Sāmoan name for sweepers.

Copper sweeper, *Pempheris otaitensis* (22 cm/9 in.)
manifi

DRUMMERS (KYPHOSIDAE)

The drummers are deep-bodied fishes with small heads and mouths. Also called sea "chubs" or "rudderfish," they are often observed swimming high above the bottom. Although drummers are omnivorous, benthic algae make up the greatest proportion of their diet.

Two species are known from the archipelago. It is illegal to fish and keep drummers smaller than than 20 cm (7.9 in.) in the western islands of Sāmoa. There is no general Sāmoan name for drummers.

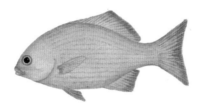

Topsail drummer, *Kyphosus cinerascens* (45 cm/18 in.)
nanue, **matāmutu**, **mutumutu** (Manuʻa Islands)

BUTTERFLYFISHES (CHAETODONTIDAE)

Butterflyfishes have highly compressed bodies and are known for their striking color patterns and graceful appearance as they swim. Their family name, Chaetodont, refers to their bristle-like teeth. Although they appear delicate, most species are actually very robust. Some members of this family are called "bannerfishes."

Most butterflyfish are diurnal and feed on individual coral polyps. Once they feed on an individual polyp, the colonial neighboring polyps withdraw and the butterflyfish must move on to another segment of coral whose polyps are exposed. Because of this restricted diet, most butterflyfishes do not survive well in a home aquarium.

Scientists believe that, unlike fish species whose sex

changes from female to male as they mature (e.g., wrasses), individual butterflyfishes remain the same sex from birth until maturity, at which time most species maintain pair bonds, perhaps for life. The pairs appear almost inseparable, and most species patrol their territories fearlessly against trespassers.

Butterflyfish coloration is usually the same in juveniles as in adults (compare with angelfish, whose young look very different from the adult). At night, many assume a less dramatic nocturnal color pattern.

Thirty species are known from the archipelago. The general Sāmoan name for butterflyfishes is **tifitifi**.

Threadfin butterflyfish, *Chaetodon auriga* (20 cm/8 in.)
ai'u, i'usamasama

Bennett's butterflyfish, *Chaetodon bennetti* (18 cm/7 in.)
tifitifi lega

Speckled butterflyfish, *Chaetodon citrinellus* (11 cm/4 in.)
tifitifi moamanu (American Sāmoa), **tifitifi muamai**
(Western Sāmoa)

Saddled butterflyfish, *Chaetodon ephippium* (23 cm/9 in.)
tifitifi tuauli

Lined butterflyfish, *Chaetodon lineolatus* (30 cm/12 in.)
tifitifi lauiʻa

Racoon butterflyfish, *Chaetodon lunula* (20 cm/8 in.)
tifitifi laumea

Ornate butterflyfish, *Chaetodon ornatissimus* (19 cm/7 in.)
tifitifi 'ava'ava

Dot-and-dash butterflyfish, *Chaetodon pelewensis* (12.5
cm/5 in.)
tifitifi tusiloloa

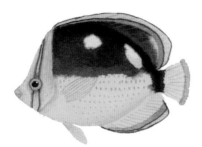

Fourspot butterflyfish, *Chaetodon quadrimaculatus* (12.6 cm/5 in.)
tifitifi segasega

Latticed butterflyfish, *Chaetodon rafflesi* (15 cm/6 in.)
tifitifi pule

Reticulated butterflyfish, *Chaetodon reticulatus* (16 cm/6 in.)
tifitifi maono

Chevroned butterflyfish, *Chaetodon trifascialis* (18 cm/7 in.)
tifitifi saeʻu

Redfin butterflyfish, *Chaetodon trifasciatus* (15 cm/6 in.)
tifitifi manifi
This species is frequently observed in pairs.

Pacific double-saddle butterflyfish, *Chaetodon ulietensis*
(15 cm/6 in.)
tifitifi gutuʻuli

Teardrop butterflyfish, *Chaetodon unimaculatus* (20 cm/8 in.)
tifitifi pulesama

Vagabond butterflyfish, *Chaetodon vagabundus* (15.6 cm/6 in.)
tifitifi matapua'a

Forcepfish, *Forcipiger flavissimus* (22 cm/9 in.)
gutumanu
Compare with longnose butterflyfish (*Forcipiger longirostris*).

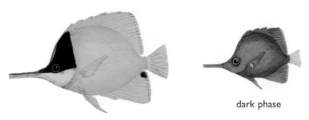

dark phase

Longnose butterflyfish, *Forcipiger longirostris* (22 cm/9 in.)
gutumanu
The longnose butterflyfish has an extremely elongated snout.
The tweezer-shaped snout, with its tiny mouth at the end,
allows the fish to pluck food from crevices on the reef that
are not accessible to other fish.

Pennant bannerfish, *Heniochus chrysostomus* (16 cm/6 in.)
laulaufau laumea

Humphead bannerfish, *Heniochus varius* (18 cm/7 in.)
laulaufau laumea

Thompson's butterflyfish, *Hemitaurichthys thompsonii*
(18.5 cm/7 in.)
This butterflyfish is abundant at Rose Atoll, uncommon elsewhere.

ANGELFISHES (POMACANTHIDAE)

Angelfishes are considered one of the most beautiful
and charismatic groups of fishes. They are closely related to
butterflyfishes and were classified in the same family until
recently. In addition to their deeply compressed bodies, they
have long spines at the corner of the gill flaps (preopercles),
which are not present on butterflyfishes. Unlike butterflyfishes, some species of angelfish exhibit dramatically different and striking juvenile color patterns compared with those
of the adult.

The male angelfish maintains a territory containing a
harem of females. Angelfishes are omnivorous, eating a variety of food items ranging from algae to benthic invertebrates. Some species are capable of producing powerful
drumming sounds that can startle the unwary snorkeler floating by. The shy but curious nature of angelfishes has made
them easy targets for thoughtless and greedy fish collectors.

Angelfishes may be observed singly or in large groups.
Some species are found in shallow water, while others occur
only in deep water. If you spot an angelfish that decides to
duck into a coral crevice to hide, wait patiently and it will
probably come out to take a peek at you.

Eleven species are known from the archipelago. The general Sāmoan name for angelfishes, **tu'u'u**, is the same as for damselfishes.

Three-spot angelfish, *Apolemichthys trimaculatus* (25 cm/10 in.)

Bicolor angelfish, *Centropyge bicolor* (15 cm/6 in.)
tu'u'u matamalū

Two-spined angelfish, *Centropyge bispinosus* (10 cm/4 in.)
tu'u'u alomu

Lemonpeel angelfish, *Centropyge flavissimus* (14 cm/5.5 in.)
tuʻuʻu sama, tuʻuʻu lega
This angelfish is similar in appearance to the three-spot angelfish, above.

Flame angelfish, *Centropyge loriculus* (10 cm/4 in.)
tuʻuʻu tusiuli
This species is most commonly observed around Swains Island.

juvenile

Emperor angelfish, *Pomacanthus imperator* (38 cm/15 in.)
tuʻuʻu vaolo (juvenile, American Sāmoa), **tuʻuʻu nuanua** (Western Sāmoa), **tuʻuʻu moana** (adult)

The juvenile of this species has an unusual bulls-eye pattern that changes to striping at maturity.

Regal angelfish, *Pygoplites diacanthus* (26 cm/10 in.)
tu'u'u moana

Interbreeding

Generally speaking, by definition, different species do not interbreed. Nevertheless, some species of angelfish interbreed with butterflyfishes, creating puzzling hybrids. Occasionally, the confusing hybrids will also breed, creating even more unusual and impossible-to-identify young.

DAMSELFISHES (POMACENTRIDAE)

Damselfishes are one of the most abundant families of coral reef fishes. Approximately 320 species occur worldwide; 27 are known from the archipelago. Likewise, damselfishes are also the most abundant family of fishes observed in most reef habitats around the archipelago (Green

1996). They have compressed bodies and range in shape from elongate to circular. This large family displays diverse preferences toward habitats, food, and behavior. The drab-colored species feed mostly on algae, while others feed on plankton. Most are territorial. If you swim over the reef, particularly "antler" coral (*Acropora* sp.), you will often see numerous damselfishes looking up and charting your progress over their individual territories. Though generally small, damselfishes have been known to nip humans passing by.

Among the most easily agitated or nervous of the territorial damselfishes are the "farmerfishes" of the genera *Stegastes*. Farmerfishes live colonially among dead branches of antler coral, where they tend their gardens of algae. They actually encourage the growth of their favored filamentous algae by weeding out other, less desirable species.

One of the well-known members of the damselfish family is the clownfish. Clownfish occur exclusively with anemones, gaining protection by hiding in the anemone's stinging tentacles. You will never find a clownfish without an anemone somewhere nearby. A special mucous coating protects the fish from stings from its anemone host. Clownfish are "proteandrous hermaphrodites." All but one individual of the group matures to be male. One of them changes sex and becomes female. If she dies, the most dominant male will change sex and take her place.

While both sexes of damselfishes usually have the same coloration, males have the ability to display distinctive courtship colors and moods by, for example, "flashing" stripes on and off. Juveniles of some species differ in coloration from adults.

Twenty-seven species are known from the archipelago. The general name for damselfishes is **tuʻuʻu** (the same as for angelfishes).

Banded sergeant, *Abudefduf septemfasciatus* (19 cm/7 in.)
mutu ("sliced," as in bread)

Indo-Pacific sergeant, *Abudefduf vaigiensis* (20 cm/8 in.)
mamo

Orange-fin anemonefish, *Amphiprion chrysopterus* (16 cm/6 in.)
tu'u'u lumane

Clark's anemonefish, *Amphiprion clarkii* (13 cm/5 in.)
tu'u'u

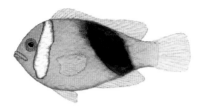

Red-and-black anemonefish, *Amphiprion melanopus* (12 cm/5 in.)
tu'u'u lumane

Midget chromis, *Chromis acares* (5.5 cm/2 in.)
tu'u'u fō
This is a dominant species at both Rose Atoll and Swains Island.

Ambon chromis, *Chromis amboinensis* (8 cm/3 in.)
tu'u'u palevai

Half-and-half chromis, *Chromis iomelas* (7 cm/3 in.)
tu'u'u i'usina

Bicolor chromis, *Chromis margaritifer* (8.5 cm/3.3 in.)
tu'u'u i'usina

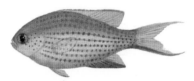

Vanderbilt's chromis, *Chromis vanderbilti* (6 cm/2 in.)
tu'u'u fō

Blue-green chromis, *Chromis viridis* (9 cm/4 in.)
i'alanumoana, tu'u'u segasega
Large aggregations of this small, iridescent fish can be
observed hovering above thickets of branching corals in
sheltered areas of the reef. It is one of the most abundant
fishes of lagoon habitats.

Pale-tail chromis, *Chromis xanthura* (15 cm/6 in.)
tu'u'u i'usina

Blueline demoiselle, *Chrysiptera caeruleolineatus* (6.5 cm/2.5 in.)

female

male

Blue devil, *Chrysiptera cyanea* (8.5 cm/3.3 in.)
tuʻuʻu moʻo, vaiuli sama
This is one of the most abundant species observed in reef flat and lagoon habitats.

Grey damsel, *Chrysiptera glauca* (11 cm/4 in.)
This is an abundant species observed in reef flat and lagoon habitats.

dark phase

Surge demoiselle, *Chrysiptera leucopoma* (8.5 cm/3.3 in.)
tu'u'u tulisegasega (blue and yellow phase), **tu'u'u alamu**
(brown phase)

South Seas demoiselle, *Chrysiptera taupou* (8.5 cm/3.3 in.)
This species is similar to *C. cyanea,* above.

Humbug dascyllus, *Dascyllus aruanus* (8.5 cm/3.3 in.)
mamo
This also is one of the most abundant species observed in
reef flat and lagoon habitats.

Three-spot dascyllus, *Dascyllus trimaculatus* (13 cm/5 in.)
tu'u'u pulelua

Dick's damsel, *Plectroglyphidodon dickii* (11 cm/4 in.)
tu'u'u i'usina
This species is most likely to be encountered in the waters of
Swains Island.

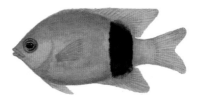

Johnston damsel, *Plectroglyphidodon johnstonianus* (10
cm/4 in.)
tu'u'u uuli
This species is similar to Dick's damsel, above, but the dark
band is usually wider and the caudal fin yellow or tan, not
white. Some individuals lack the dark band altogether.

Jewel damsel, *Plectroglyphidodon lacrymatus* (11 cm/4 in.)
tu'u'u lau, i'usamasama

Charcoal damsel, *Pomacentrus brachialis* (11 cm/4 in.)
tu'u'u faga

Neon damsel, *Pomacentrus coelestis* (10 cm/4 in.)
tu'u'u segasega

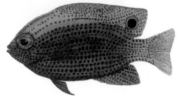

Princess damsel, *Pomacentrus vaiuli* (10 cm/4 in.)
tu'u'u vaiuli
This species' scientific name most likely comes from its
Sāmoan name.

Whitebar Gregory, *Stegastes albifasciatus* (12 cm/5 in.)
tu'u'u pa, ulavapua
This is one of the most abundant fishes in lagoon habitats.

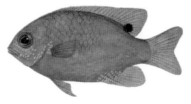

Dusky Gregory, or farmerfish, *Stegastes nigricans* (15 cm/6 in.)
tu'u'u moi

HAWKFISHES (CIRRHITIDAE)

The hawkfishes have typical fish-shaped bodies with numerous short filaments at the tips of their dorsal spines. Most species have large green or black eyes rimmed with gold. Hawkfishes are usually found resting on the bottom or perched horizontally on a branch of coral. They have thick pectoral fins that can be used to wedge themselves into place. Some species have large lips and resemble miniature groupers. Almost all members of this family are carnivorous "lie-in-wait" predators that ambush unsuspecting prey that passes nearby. Some males appear to be territorial and maintain a harem of females.

At least eight species occur in the archipelago. There is no general Sāmoan name for this family.

Stocky hawkfish, *Cirrhitus pinnulatus* (28 cm/11 in.)
ulutu'i

Arc-eye hawkfish, *Paracirrhites arcatus* (14 cm/6 in.)
lausiva

Blackside hawkfish, *Parachirrhites forsteri* (22.5 cm/9 in.)
lausiva

MULLETS (MUGILIDAE)

Mullets are silvery gray and are shaped somewhat like a submarine. They are usually seen in schools swimming quickly at midwater level. Mullets are a feast for predators such as barracudas and jacks.

Seven species are known from the archipelago. Mullets are an important food fish in many areas, and each size class has its own Sāmoan name. It is illegal to fish and keep mullets less than 20 cm (7.9 in.) in the western islands of Sāmoa. The general Sāmoan name for mullet (21–40 cm) is **'anae**. Other size classes are named as follows: **moi** (< 5 cm), **poi** (5–8 cm), **Aua** (9–12 cm), **fuafua** (13–15 cm), **popoto** or **manase** (16–20 cm), and **afomatua** (> 40 cm).

Diamond-scale mullet (also known as squaretail mullet),
Ellochelon vaigiensis (55 cm/22 in.)

fuitogo (< 10 cm), **ʻafa** (10–25 cm), **ʻanaeafa** (> 25 cm).
Look for the darkly colored pectoral fin (juvenile characteristic) held high against the body. The genus of this species
was formerly *Liza*.

BARRACUDAS (SPHYRAENIDAE)
The barracuda is well known for its large mouth filled
with long, very sharp teeth. The two dorsal fins are widely
separated, and the caudal fin is forked. The elongate body is
silvery, sometimes with dark markings or bars. The head is
also long, with an extended lower jaw that adds to the fierce
appearance.

Although barracudas can occur in large schools, the
great barracuda is frequently solitary. It may be attracted to
divers and snorkelers, and may come up close for a look—
not a bite—out of curiosity. Its dangerous reputation is
unwarranted. Attacks on humans have usually been cases of
mistaken identity that took place in very murky water or
because of incitement from being impaled during spearfishing.

Five species are known from the archipelago. They are
a delicious food fish, but large barracudas have been known
to be ciguatoxic. Barracudas are generally called **sapatū**,
though the species illustrated is known by another name.

Great barracuda, *Sphyraena barracuda* (170 cm/67 in.)
saosao

WRASSES (LABRIDAE)
Wrasses are probably the most diverse family of reef

fishes in both size and shape. They vary in length from a few centimeters, e.g., redlip cleaner wrasse, *Labroides pectoralis,* 9 cm (3.5 in.), to the huge humphead maori wrasse, *Cheilinus undulatus,* which reaches 2.3 m (7 ft.). Most species exhibit fascinating color patterns that vary with age and sex. Medleys of colors are juxtaposed in an endless combination of spots, dots, squiggles, and stripes.

Many species have a juvenile phase, followed by an initial phase, and finally a terminal phase (wrasses at this stage are also called "supermales"). Color patterns often change dramatically as the fish matures from one phase to another. To add to the confusion, all species are believed to be protogynous hermaphrodites: females have the capacity to become males. In other words, in the terminal phase, some wrasses who are male were previously female. Other individuals are born male and stay male.

Most wrasses are carnivorous and feed on benthic invertebrates such as crabs and sea urchins. Wrasses are diurnal, and some species are known to sleep beneath the sand. It is suspected that wrasses need a lot of sleep, as they are the first group of fishes to sleep at night and the last to wake up in the morning.

Those of the genus *Labroides*, and the young of some other species, feed primarily on ectoparasites located on the bodies of other fish. They are thus known as "cleaner wrasses." Cleaner wrasses occupy an area on the reef that many other fish visit for a cleaning. Cleaner wrasses will even search for parasites from within the mouths of large predatory fishes such as barracudas. Because cleaner wrasses serve this important function, predatory fishes will not harm them.

At least 64 species are known from the archipelago. It is illegal to fish for and keep wrasses less than 20 cm (7.9 in.) in the western islands of Sāmoa. The general Sāmoan name for wrasses is **sugale**.

Spotted wrasse, *Anampses meleagrides* (21 cm/8 in.)
sugale tatanu (initial color phase)
Males have squiggly blue lines on the face, vertical blue
lines on each scale, and blue spots on the caudal peduncle.

Yellowbreasted wrasse, *Anampses twistii* (18 cm/7 in.)
sugale tatanu

juvenile

Axilspot hogfish, *Bodianus axillaris* (20 cm/8 in.)
sugale vaolo

Floral maori wrasse, *Cheilinus chlorourus* (36 cm/14 in.)
lalafi matapua'a

Cheeklined maori wrasse, *Oxycheilinus digrammus* (formerly genus *Cheilinus*) (30 cm/12 in.)
lalafi gutu'umi

Redbreasted maori wrasse, *Cheilinus fasciatus* (36 cm/14 in.)
lalafi pulepule

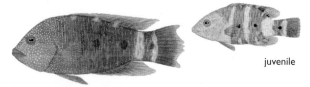

juvenile

Tripletail maori wrasse, *Cheilinus trilobatus* (40 cm/16 in.)
lalafi matamūmū

juvenile

Humphead maori wrasse, *Cheilinus undulatus* (229 cm/90 in.)
lalafi (< 30 cm), **tagafa** (30–75 cm), and **malakea** (> 75 cm)

Ringtail maori wrasse, *Oxycheilinus unifasciatus* (formerly in the genus *Cheilinus*) (46 cm/18 in.)
lalafi

Scott's wrasse, *Cirrhilabrus scottorum* (13 cm/5 in.)

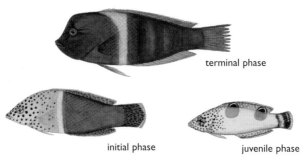

terminal phase

initial phase   juvenile phase

Clown coris, *Coris aygula* (100 cm/39 in.)
sugale uluto'i

juvenile phase

Yellowtail coris, *Coris gaimard* (40 cm/16 in.)
sugale mūmū, sugale tala'ula

terminal phase

Slingjaw wrasse, *Epibulus insidiator* (35 cm/14 in.)
lapega (American Sāmoa), **si'umutu** (Western Sāmoa),
lalafi tua'au
To capture its prey, the slingjaw wrasse can extend its jaw to
a distance equal to one-third of its body length; in its initial
phase, it is solid lemon yellow.

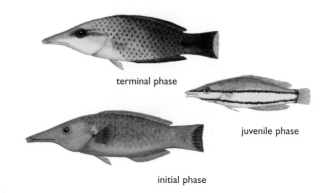

terminal phase

juvenile phase

initial phase

Bird wrasse, *Gomphosus varius* (32 cm/13 in.)
gutusi'o, gutu'umi, sugale lupe

juvenile phase

Checkerboard wrasse, *Halichoeres hortulanus* (27 cm/11 in.)
sugale a'au, sugale pagota, ifigi

initial phase

Dusky wrasse, *Halichoeres marginatus* (17 cm/7 in.)
sugale lalafi

initial phase

Tailspot wrasse, *Halichoeres melanurus* (10.5 cm/4 in.)

initial phase

Nebulous wrasse, *Halichoeres nebulosus* (12 cm/5 in.)

Ornate wrasse, *Halichoeres ornatissimus* (15 cm/6 in.)

Threespot wrasse, *Halichoeres trimaculatus* (20 cm/8 in.)
lape, **sugale pagota**

juvenile phase

Barred thicklip, *Hemigymnus fasciatus* (75 cm/30 in.)
sugale gutumafia

juvenile phase

Blackeye thicklip, *Hemigymnus melapterus* (90 cm/35 in.)
sugale laugutu, **sugale uli**, **sugale aloa**, **sugale lupe**

Bicolor cleaner wrasse, *Labroides bicolor* (14 cm/6 in.)
sugale iʻusina

Striped cleaner wrasse, *Labroides dimidiatus* (11.5 cm/4.5 in.)
sugale moʻotai

Redlip cleaner wrasse, *Labroides rubrolabiatus* (9 cm/4 in.)

initial phase

Yellowback tubelip, *Labropsis xanthonota* (13 cm/5 in.)

initial phase

Blackspotted wrasse, *Macropharyngodon meleagris* (15 cm/6 in.)
sugale puletasi

juvenile phase

Rockmover wrasse, *Novaculichthys taeniourus* (30 cm/12 in.)
sugale laʻo (juvenile), **sugale tāili** (adult), **sugale gasufi**

Sixstripe wrasse, *Pseudocheilinus hexataenia* (7.5 cm/3 in.)
sugale tusitusi (American Sāmoa), **sugale manifi** (Western Sāmoa)

initial phase

Bluelined wrasse, *Stethojulis bandanensis* (12.5 cm/5 in.)
lape aʻau

initial phase

Stripebelly wrasse, *Stethojulis strigiventer* (15 cm/6 in.)
lape aʻau

Six-barred wrasse, *Thalassoma hardwicke* (18 cm/7 in.)
sugale a'au, lape ele'ele
The six-barred wrasse is a particularly abundant and characteristic fish of lagoon habitats around the islands of the archipelago.

Sunset wrasse, *Thalassoma lutescens* (24.7 cm/9.7 in.)
sugale samasama ("wrasse-yellow")

Surge wrasse, *Thalassoma purpureum* (43 cm/17 in.)
uloulo gatala (initial phase), **patagaloa** (terminal phase)

initial phase

Fivestripe wrasse, *Thalassoma quinquevittatum* (17 cm/7 in.)
lape moana

Ladder wrasse, *Thalassoma trilobatum* (30 cm/12 in.)

PARROTFISHES (SCARIDAE)

Parrotfishes are so named because of their gaudy, yet beautiful colors and because their top and bottom teeth are fused in the shape of a beak. This family of fishes evolved from the wrasse family. Nevertheless, parrotfishes are primarily herbivorous, not carnivorous like wrasses. They feed on algae that they nip from the outside layer of corals. Although parrotfishes are considered herbivores because most of the food found in their stomachs is plant material (algae), they inadvertently consume some animal matter when browsing on the coral and its polyps. These small pieces of coral are devoured and later excreted as sand. In this way, parrotfishes are also major producers of the sandy sediment of our tropical oceans. The telltale markings from the bites of parrotfishes are easy to see on large *Porites* coral heads. They appear as small scrape marks about a few centimeters long.

There is great variation in parrotfish size. In general, they are large, bulky fishes, and some species are huge. For example, the bumphead parrotfish is known to reach 120 cm (almost 4 ft.) and weigh 46 kg (over 100 lbs.). The scales of the bumphead parrotfish are so thick that they can allegedly deflect a metal spear at point-blank range.

Like wrasses, parrotfishes have three distinct phases: juvenile, initial, and terminal (supermale). The juvenile and initial phases usually comprise both males and females. Females are able to change their sex to become "terminal" males. At the same time, there is a dramatic change in coloration. Because of these uniquely colored sexes, scientists initially determined, incorrectly, that the color phases were two different species. Terminal males of some species establish territories and maintain harems of females.

Some parrotfishes fabricate a balloon-shaped "sleeping bag" of mucus around themselves before going to sleep. Scientists speculate that this serves the purpose of hiding the parrotfish's scent from potential predators, but no one knows for sure. Unfortunately, even with a mucous cocoon, sleeping parrotfishes are easy prey for spear fishermen.

Twenty-two species are known from the archipelago. Parrotfishes are an important food fish, but it is illegal to fish for and keep parrotfishes less than 20 cm (7.9 in.) in the western islands of Sāmoa. The general Sāmoan name for parrotfishes varies with size and coloration. Small to medium-sized parrotfishes (< 20 cm) are called **fuga**, while larger individuals are called **laea** (20–50 cm) or **galo** (> 50 cm). Reddish brown fish are referred to as **fugamea**, and greenish blue fish are called **fugausi**.

Bumphead parrotfish, *Bolbometopon muricatum* (120 cm/47 in.)
uluto'i ("head like an adze"; < 20 cm), **laea uluto'i** (20–50 cm), **galo uluto'i** (> 50 cm)

initial phase

Stareye parrotfish, *Calotomus carolinus* (50 cm/20 in.)

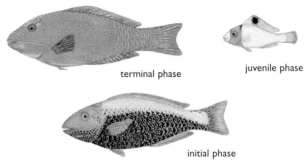

terminal phase

juvenile phase

initial phase

Bicolor parrotfish, *Cetoscarus bicolor* (80 cm/31 in.)
fuga sina (juvenile), **mamanu** (initial phase < 25 cm), **laea
mamanu** (initial phase > 25 cm), **laea usi** (terminal phase)

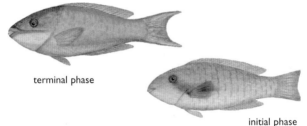

terminal phase

initial phase

Whitespot parrotfish, *Scarus forsteni* (55 cm/22 in.)

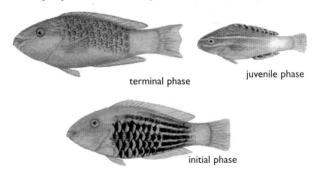

terminal phase

juvenile phase

initial phase

Bridled parrotfish, *Scarus frenatus* (47 cm/19 in.)
laea mea (initial phase), **laea si'umoana** (terminal phase)

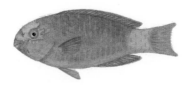

Reefcrest parrotfish, *Chlorurus frontalis* (formerly genus *Scarus*) (50 cm/20 in.)

terminal phase

initial phase

Globehead parrotfish, *Scarus globiceps* (27 cm/11 in.)

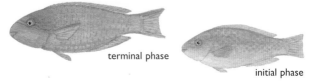

Steephead parrotfish, *Chlorurus microrhinos* (formerly genus *Scarus*) (70 cm/28 in.)
fugausi (< 25 cm), **laea** (25–40 cm), **ulumato** (41–50 cm), **galo** (> 50 cm)
This species, in particular, grinds a lot of coral into sand.

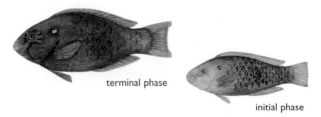

terminal phase

initial phase

Swarthy parrotfish, *Scarus niger* (35 cm/14 in.)
fuga pala (< 25 cm), **laea pala** (> 25 cm)

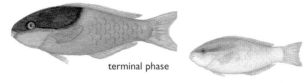

Egghead parrotfish, *Scarus oviceps* (30 cm/12 in.)
fuga alosina (initial phase), **laea tuavela** (terminal phase)

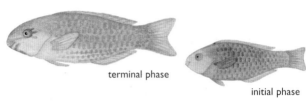

Palenose parrotfish, *Scarus psittacus* (30 cm/12 in.)
fuga matapua'a (< 15 cm), **fugausi matapua'a** (15–25 cm), **laea matapua'a** (> 25 cm)

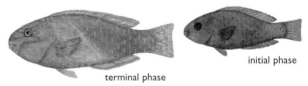

Redtail parrotfish, *Chlorurus japanensis* (formerly *Scarus pyrrhurus*) (30 cm/12 in.)
fuga si'umu (initial phase), **laea ulusama** (terminal phase)

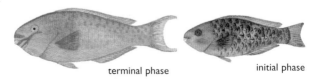

Ember parrotfish, *Scarus rubroviolaceus* (70 cm/28 in.)
laea mea (initial phase), **laea mala** (terminal phase)

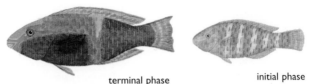

terminal phase

initial phase

Schlegel's parrotfish, *Scarus schlegeli* (38 cm/15 in.)
fuga matapua'a (initial phase), **laea tusi** (terminal phase)

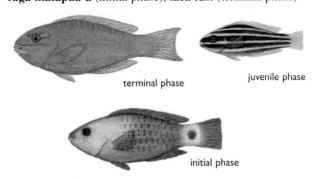

terminal phase

juvenile phase

initial phase

Bullethead parrotfish, *Chlorurus sordidus* (formerly genus
Scarus) (40 cm/16 in.)
fuga gutumū (initial phase), **fugausi tuavela** or **laea
tuavela** (terminal phase)

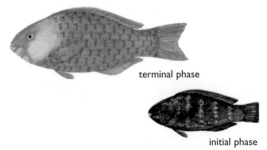

terminal phase

initial phase

Greensnout parrotfish, *Scarus spinus* (30 cm/12 in.)
fua a'au

Overfishing

Parrotfishes are an important food fish in the Pacific. But their numbers are declining precipitously in many areas. Because sleeping parrotfishes are especially unwary, nighttime spearfishermen equipped with lights and scuba gear are easily able to kill large numbers. Parrotfishes in Guam, for example, have almost been wiped out. In the Sāmoan Archipelago, too, this type of spearfishing is resulting in overfishing of parrotfishes and other species and a dangerous decrease in the number of mature, breeding individuals. If action is not taken now to stop such overfishing, this fishery will be destroyed in a short period of time.

In addition to being an important food fish, parrotfishes are directly responsible for protecting the islands from erosion during storms. How so? Parrotfishes are living coral-processing factories. They scrape, bite, crush, and swallow the coral, extract the nutrients during digestion, and then pass the sandy waste back into the water. An amazing amount of sandy material is created this way. A single adult parrotfish (e.g., steephead parrotfish, *Chlorurus microrhinos*) may excrete more than a ton of sand each year. This sand builds up beaches and shorelines, thereby helping to protect the islands.

🐟🐟🐟🐟🐟🐟🐟🐟🐟🐟🐟🐟🐟🐟🐟🐟🐟

Sand Mining

Coral reefs and beaches are nature's way of protecting islands from the force of the ocean. The volume and location of sand on beaches depends on the rate and nature of its origin. The rate of sand deposition depends on the rate at which coral and other organisms, e.g., the coralline algal plant Halimeda sp. (see Marine Invertebrates), break down.

During quiet periods, the sand is moved around gently by waves and currents. In contrast, during storms, more sand is moved, and the results can be dramatic. If the beaches are robust, they can withstand the fury of waves, larger storms, and even cyclones. The sand and coral rubble may be moved offshore by the storm, but it will eventually return and build up the shoreline again. But if people take all the sand and coral rubble away for their own purposes, there is not enough left along the shoreline when a storm comes, and consequently the very land itself will be eroded instead of the beach. Exposed "beach rocks" and steep drop-offs along shorelines show where coral sand beaches used to be (e.g., Faagalu and Fatu ma Futi in American Sāmoa).

Sea grass beds, an important habitat and source of food for herbivorous green sea turtles (Chelonia mydas), also help to stabilize drifting sand with their extensive root systems.

If you live in the Sāmoa Islands, ask older members of your village if the beach has changed in their lifetime.

🐟🐟🐟🐟🐟🐟🐟🐟🐟🐟🐟🐟🐟🐟🐟🐟🐟

SANDPERCHES (PINGUIPEDIDAE)

These fishes have elongate, cigar-shaped bodies that are slightly compressed. Their heads end in long, pointed snouts with pouting mouths. Sandperches do exactly what their name implies: they live perched in open areas of the bottom, on sand or rubble. They rest on the substrate by propping themselves up using their pectoral fins. Their eyes are directed upward, and they are often confused with the lizardfishes (Synodontidae). All sandperches are carnivorous and feed primarily on benthic crustaceans. Some species occur at an incredible depth of 360 m (1,181 ft.).

Three species are known from the archipelago. All three have the same Sāmoan name, **ta'oto** ("to lie flat").

Latticed sandperch, *Parapercis clathrata* (17.5 cm/6.9 in.)
ta'oto

Speckled sandperch, *Parapercis hexophtalma* (23 cm/9 in.)
ta'oto

Spotted sandperch, *Parapercis millepunctata* (18 cm/7 in.)
ta'oto

BLENNIES (BLENNIIDAE)

Blennies are typically small, scaleless, elongate fishes with bluntly shaped heads. They have a single long dorsal fin. Blennies are generally cryptic bottom-dwellers, and many species look similar. This can make identification in the field difficult. Some have unusual crests or fringe decorating their heads. Others live close to shore on rocky substrate occurring in the wave-swept intertidal zone. Some species even leap from one rock or tide pool to another, hence the common name "rockskipper." The majority of blennies are herbivores, though some are carnivorous (see inset).

Blennies do not exhibit traditional romance and courtship behavior. Instead, it is the female that romances the male, and the male that tends and defends their eggs.

At least 28 species are known from the archipelago. The general Sāmoan name for blennies is **mano'o**.

Mimic blenny, *Aspidontis taeniatus* (11.5 cm/4.5 in.)
mano'o mo'o, mo'otai
This blenny mimics the striped cleaner wrasse (*Labroides dimidiatus*). See inset, p. 114.

Bicolor blenny, *Ecsenius bicolor* (11 cm/4 in.)
mano'o i'usama

Shortbodied blenny, *Exallias brevis* (14.5 cm/5.7 in.)
mano'o lau, mano'o gatala
Some blenny species, such as this one, are prolific repro-
ducers, laying an estimated 200,000 to 300,000 eggs per
year. Males have bright red spots; females (illustrated),
brown.

Rippled rockskipper, *Istiblennius edentulus* (17 cm/7 in.)
Rippled rockskippers are common along rocky shores, living in the wave-washed intertidal zone, where they can be
observed jumping from rock to rock or tide pool to tide
pool.

Yellowtail fangblenny, *Meiacanthus atrodorsalis* (11 cm/4
in.)
mano'o si'umaga

Mimicry

The mimic blenny (Aspidontus taeniatus) imitates the cleaner wrasse (Labroides dimidiatus) in almost every aspect, from coloration to behavior. Its mimicry allows it to get close to any unsuspecting fish that has stopped by to be groomed (of its ectoparasites) by what it thinks is a cleaner wrasse. The deceptive blenny zooms in, quickly takes a bite with two very large, sharp teeth, and disappears before the victim realizes its mistake.

GOBIES (GOBIIDAE)

Gobies are the largest family of marine fishes in the world. There are approximately 1,600 species worldwide, with 1,200 of those occurring in the Indo-Pacific. There are substantial variations in size, color, and form, and most species are carnivorous. Gobies are usually elongate, with two dorsal fins.

Most gobies hover on, or just over, the substrate. Because they often disappear at the first sign of danger, they are frequently overlooked. Some species rest on the branches of coral, while others are common in caves or crevices in the outer reef. Many species occupy burrows that they share with only a single shrimp (see inset). Other species are very small and difficult to observe. In fact, the tiniest vertebrate (animal with a backbone) in the world is the Philippine goby (*Pandaka pygmaea*), which, when full-grown, measures only 1 cm (less than one-half inch).

Many species of goby living in the waters surrounding the archipelago have yet to be identified, because of confusing taxonomy or lack of descriptions in scientific literature.

One hundred species are known from the archipelago, though most are never observed. While gobies are usually too small for human consumption, they are an important part of the reef food web because of their abundance. The general Sāmoan name for gobies is the same as that for blennies, **mano'o**.

Fourbar goby, *Gobiodon citrinus* (6.6 cm/2.6 in.)
mano'o ulutu'i, moemimi

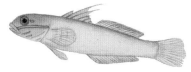

Blueband goby, *Valenciennea strigata* (18 cm/7 in.)
mano'o sina

Symbiosis

The lives of some reef animals are closely connected with, or dependent upon, the lives of another species. This relationship is generally known as "symbiosis." There are three basic forms of symbiosis: (1) mutualism, in which both organisms depend upon each other; (2) commensalism, in which one organism (the commensal) obtains some benefit from another organism without causing harm to it; and (3) parasitism, in which one organism (the parasite) benefits to the detriment of another.

Mutualism

Some species of goby (called shrimp gobies) enjoy a mutualistic relationship with certain species of shrimp by sharing a burrow. The shrimp's job is to build and maintain the burrow, while the goby's duty is to guard the entrance of the burrow from predators. The smaller shrimp probably also benefits from leftover morsels of the goby's meals.

Commensalism

An example of a commensal relationship is that between a clownfish and its anemone host. The clownfish (the commensal), which is immune to the anemone's stinging tentacles, spends most of its time hiding within the safety of these arms. The anemone does not really benefit from this relationship, although it may receive a few leftover scraps of food from the clownfish.

Parasitism

There is a strange, sometimes parasitic relationship between some sea cucumbers and members of the pearlfish (Carapidae) family. The tiny, finless, extremely elongate fish lives inside the anal opening of the sea cucumber (e.g., leopard sea cucumber), coming out only to forage for food. At the first approach of danger, the fish retreats quickly into its hideaway by backing in tail first. Some parasitic pearlfish feed on the gills and sexual organs of their host as well.

DARTFISHES (MICRODESMIDAE)

The dartfishes, also known as "wormfishes" and "hovergobies," are related to the gobies. They have elongate bodies and usually heavy, extended lower jaws. Unlike gobies, dartfishes have divided dorsal fins.

Most species of this secretive fish occur over sandy or muddy bottom habitats and retreat into their burrows if disturbed.

Nine species are known from the archipelago. The general Sāmoan name for dartfishes is **mano'o ui**.

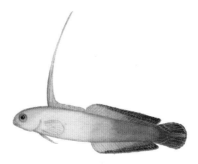

Fire dartfish, *Nemateleotris magnifica* (8 cm/3 in.)
mano'o sugale

Two-tone dartfish, *Ptereleotris evides* (13.8 cm/5.4 in.)
ma'ulu

Zebra dartfish, *Ptereleotris zebra* (12 cm/5 in.)

SURGEONFISHES (ACANTHURIDAE)

The surgeonfishes are so named because of the sharp, knifelike spine (or spines) located just in front of the caudal fin. They are also known generally as "tangs" or "doctor-fishes." The surgeonfish family consists of 72 species, and is characterized by a compressed body shape, eyes located high on the head, and a small mouth. Most surgeonfishes are her-bivorous and graze on benthic algae, but a few carnivorous species feed on zooplankton.

While some species are colorful and easy to identify, others are brown, and one must look carefully at specific fea-tures to identify the fish in the field (e.g., compare lined bristletooth and brown surgeonfish). One group of surgeon-fishes (genus *Naso*) are called "unicornfishes" because of various-sized hornlike extensions on the forehead of adults. The reason for the horn is unknown, but one explanation is that it makes the adult fish harder to swallow. While adults are easy to identify, juvenile unicornfishes all look similar, and identification is difficult.

The hinged spine of surgeonfishes usually lies flat against the body and is similar to the blade of a switchblade knife. When necessary, the surgeonfish raises this sharp cau-dal spine for protection and sweeps it sideways at the approaching source of danger. The spine area is sometimes brightly colored and probably serves to draw the attention of any would-be predator to its weapon. Some surgeonfishes are capable of altering their colors.

Surgeonfishes are conspicuous reef inhabitants and are most abundant on the reef crest. They are also an important food fish. It is illegal to fish for and keep surgeonfishes less than 20 cm (7.9 in.) in the western islands of Sāmoa.

Thirty-two species are known from the archipelago. The general Sāmoan name for small *Acanthurus* sp. (< 15 cm) is **pone**; larger fish are called **palagi**. *Naso* sp. are generally called **ume**, with the smallest called **ʻiliʻilia** or **umelei**.

Achilles tang, *Acanthurus achilles* (19 cm/7 in.)
maikolama, kolama, pone i'umumu
This species is most commonly observed around Rose
Atoll.

Ringtail surgeonfish, *Acanthurus blochii* (42 cm/17 in.)
This fish looks nearly black underwater except for a white
ring on the base of the caudal fin.

Eyestripe surgeonfish, *Acanthurus dussumieri* (50 cm/20
in.)

White-spotted surgeonfish, *Acanthurus guttatus* (21.5 cm/8.5 in.)
maogo

Striped surgeonfish, *Acanthurus lineatus* (38 cm/15 in., though typically half this length in the Sāmoas)
alogo

White-cheek surgeonfish, *Acanthurus nigricans* (21.3 cm/8.4 in.)

Brown surgeonfish, *Acanthurus nigrofuscus* (21 cm/8 in.)
ponepone
Brown surgeonfishes are most commonly observed around
the islands of Manuʻa; they are similar in appearance to the
goldring bristletooth (*Ctenochaetus strigosus*) and lined
bristletooth (*Ctenochaetus striatus*), below.

Orangeband surgeonfish, *Acanthurus olivaceus* (35 cm/14
in.)
pone apasama, afinamea

Convict surgeonfish, *Acanthurus triostegus* (26.3 cm/10.4
in.)
manini

The convict tang, named for its striped appearance, resembling a prison uniform, is one of the most abundant fishes observed on reef flats, occurring in large groups moving across the reef. Convict tangs may graze the algae growing on the shells of green turtles. This may benefit the turtles if the algal growth decreases the hydrodynamic efficiency of their movements through the water.

Goldring bristletooth, *Ctenochaetus strigosus* (18 cm/7 in.)
The goldring bristletooth is most commonly observed around Rose Atoll; it is similar in appearance to the brown surgeonfish (*Acanthurus nigrofuscus*) and lined bristletooth (*Ctenochaetus striatus*).

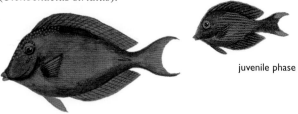

juvenile phase

Lined bristletooth, *Ctenochaetus striatus* (26 cm/10 in.)
pone (adults), **pala'ia** or **logoulia** (schooling juveniles)
The lined bristletooth is an abundant and characteristic fish of lagoon habitats around the archipelago, similar in appearance to the brown surgeonfish (*Acanthurus nigrofuscus*), above, and goldring bristletooth *Ctenochaetus strigosus,* below.

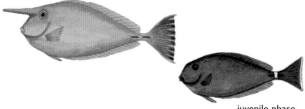

juvenile phase

Whitemargin unicornfish, *Naso annulatus* (100 cm/39 in.)

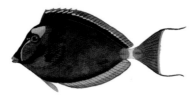

Orangespine unicornfish (also called the "lipstick tang"), *Naso lituratus* (45 cm/18 in.)
'ili'ilia (< 15 cm), **umelei** (>1 5 cm).
This fish is abundant around the islands of Manu'a.

Humpnose unicornfish, *Naso tuberosus* (60 cm/24 in.)
ume uluto'i

Bluespine unicornfish, *Naso unicornis* (70 cm/28 in.)
ume isu (**isu** means "nose")

Palette surgeonfish, *Paracanthurus hepatus* (31 cm/12 in.)

Brushtail tang, *Zebrasoma scopas* (20 cm/8 in.)
pitapita, pe'ape'a

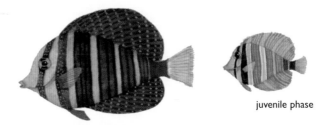

juvenile phase

Sailfin tang, *Zebrasoma veliferum* (40 cm/16 in.)

MOORISH IDOL (ZANCLIDAE)

Upon first look, you might confuse the Moorish idol with a member of the butterflyfish family (compare with bannerfish). In actuality, it is more closely related to the surgeonfishes, but lacks the scalpel-like spine at the base of the caudal fin. One of the dorsal spines on this fish is extremely long, flowing, and filamentous, and its length varies with each individual. The snout is long and pointed, with a small mouth at the end. Adults have a bony projection above the eye, which is largest in males. Moorish idols can be found both in the shallows and to depths of 180 m (591 ft.) or more. They are usually seen in small groups and are omnivorous, eating both plants and animals.

There is only one species in the family Zanclidae. The general Sāmoan names for the Moorish idol are **pe'ape'a** ("bat") and **laulaufau**.

Moorish idol, *Zanclus cornutus* (14 cm/6 in.)

RABBITFISHES (SIGANIDAE)

The rabbitfishes are also known as "spinefeet." This name comes from their venomous pelvic fin, which can inflict a painful wound. Their bodies are ovate and compressed, and the caudal peduncle (base of the caudal fin) is narrow. Rabbitfishes are diurnal, and most are herbivorous, eating large amounts of sea grass. Those seen on the reef usually occur in pairs. Rabbitfishes can change their color when at rest near the bottom.

At least three species are known from the archipelago. Most species are food fish, though none are very large. Nevertheless, it is illegal to fish and keep rabbitfishes smaller than 20 cm (7.9 in.) in the western islands of Sāmoa. The general Sāmoan name for rabbitfishes is **lō**, used when referring to large schools of juveniles.

Spiny rabbitfish, *Siganus spinus* (19 cm/7 in.)
anefe (< 5 cm), **pa ʻulu** (> 5 cm)

BILLFISHES (ISTIOPHORIDAE)

The marlins, spearfishes, and sailfishes are all well-known game fish, typically having highly elongated snouts known as "bills." These active, oceanic fishes use their bills to attack their prey, which usually comprises other fish and cephalopods (e.g., squid).

It was once common to see photographs of fishermen standing alongside captured marlins at marina docks. Today, charterboat operators and recreational fishermen are encouraged to tag and release their fish rather than continue the senseless killing of these magnificent seafarers.

Five species of billfish are known from the archipelago. The general Sāmoan name for billfishes is **saʻulā**.

Indo-Pacific blue marlin, *Makaira mazara* (500 cm/197 in.)
saʻulā oso ("long-snout jumping fish")

TUNAS AND MACKERELS (SCOMBRIDAE)

These well-known fishes are the basis of important commercial and local individual village fisheries. There are two large tuna processing businesses in American Sāmoa.

Tunas and mackerels are powerful swimmers, reaching speeds of 64 kph (40 mph), and can move long distances across the ocean. Typically, all other fish species are exothermic, but some species of tuna have body temperatures several degrees warmer than the surrounding sea.

Eleven species are known from the archipelago. There is no general Sāmoan name for the Scombrids.

Wahoo, *Acanthocybium solandri* (210 cm/83 in.)
paāla

Skipjack tuna, *Katsuwonus pelamis* (110 cm/43 in.)
atu (< 40 cm), **faolua** (40–50 cm), **ga'ogo** (> 50 cm)

Albacore tuna, *Thunnus alaluga* (150 cm/59 in.)
apakoa

Yellowfin tuna, *Thunnus albacares* (210 cm/83 in.)
asiasi (< about 18 kg), **tuʻuo** (American Sāmoa), **taʻuo** (Western Sāmoa, > about 18 kg)
The bright yellow fin color fades quickly upon death.

Bigeye tuna, *Thunnus obesus* (240 cm/94 in.)
asiasi (< about 18 kg), **tuʻuo** (American Sāmoa), **taʻuo** (Western Sāmoa, > about 18 kg)

TRIGGERFISHES (BALISTIDAE)

Triggerfishes are among the most unusual-looking fishes. They have relatively deep, compressed bodies, eyes set far back and high on their heads, and small, terminal mouths, some of which point (and pout) upward. The triggerfish is so named because it has a large dorsal spine that it can lock into position by "triggering" an adjacent smaller spine. When alarmed, triggerfishes wedge themselves headfirst into a crevice and erect their spines, thereby essentially locking themselves in place. A triggerfish will often settle down for the night in this position, with only the tail visible—sticking out from the fish's hiding place. Triggerfishes typically swim by undulating their second dorsal and anal fins, using their tails only for rapid bursts of speed.

Triggerfishes are usually solitary, and some of the smaller species are shy. Most are diurnal carnivores that eat a variety of benthic organisms such as crabs, urchins, and mollusks. They have powerful jaws capable of crushing hard-shelled prey. Some have been observed squirting powerful jets of water at sea urchins to topple them in order to get to their softer undersides. Urchins are good to eat, but only if one is very careful, as the spines are toxic and can inflict painful wounds.

Triggerfishes are one of the few fish families that exhibit parental care. In this case, the mother guards the eggs. Triggerfishes have been known to inflict painful bites on divers who wander much too close to their territory or eggs.

Triggerfishes are related to filefishes, boxfishes, puffers, and porcupinefishes. Although they are a food fish, some species may be ciguatoxic.

Fifteen species are known from the archipelago, and their general Sāmoan name is **sumu**.

Orange-lined triggerfish, *Balistapus undulatus* (30 cm/12 in.)
sumu aimaunu

Clown triggerfish, *Balistoides conspicillum* (50 cm/12 in.)
sumu papa

Titan triggerfish, *Balistoides viridescens* (75 cm/30 in.)
sumu laulau (< 20 cm), **umu** (> 20 cm)
Females guarding nests may be aggressive.

Black triggerfish, *Melichthys niger* (28.1 cm/11 in.)
sumu uli
Black triggerfishes are most commonly observed around
Swains Island.

Pinktail triggerfish, *Melichthys vidua* (35 cm/14 cm)
sumu ʻapaʻapasina, sumu siʻumūmū
The pinktail triggerfishes are most commonly observed
around Swains Island.

Redtooth triggerfish, *Odonus niger* (40 cm/16 in.)
sumu peʻa

Yellow-margin triggerfish, *Pseudobalistes flavimarginatus*
(55 cm/22 in.)
sumu laulau (< 20 cm), **umu** (> 20 cm)
Females guarding nests may be aggressive.

Whitebanded triggerfish, *Rhinecanthus aculeatus* (25
cm/10 in.)
sumu uoʻuo

Wedge-tail triggerfish, *Rhinecanthus rectangulus* (25 cm/10 in.)
sumu aloalo

Scimitar triggerfish, *Sufflamen bursa* (24 cm/9 in.)
sumu pa'epa'e

Flagtail triggerfish, *Sufflamen chrysopterus* (22 cm/9 in.)
sumu gasemoana

Bridled triggerfish, *Sufflamen fraenatus* (38 cm/15 in.)
sumu gase'ele'ele

Gilded triggerfish, *Xanthichthys auromarginatus* (22 cm/9 in.)
sumu palu
Females are dark reddish brown.

FILEFISHES (MONOCANTHIDAE)

Filefishes are closely related to triggerfishes, and the two are often mistaken for each other. They have a similar erectable spine on the head, but a more narrow, compressed body. The spine on the filefish is located over the eye, while on the triggerfish it is set farther back. Unlike triggerfishes, most filefishes are able to change their color to match their surroundings. They are also secretive, while their cousins are bolder. It is important to carefully look at their shape for identification. Most are omnivorous and feed on a great variety of plants and animals. They are also known as "leatherjackets" in some areas because of their tough, leathery skin.

At least six species are known from the archipelago. The general Sāmoan name for the filefishes is **pa'umalō**.

Unicorn leatherjacket, *Aluterus monoceros* (75 cm/30 in.)

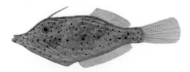

Scrawled leatherjacket, *Aluterus scriptus* (75 cm/30 in.)
ume aleva, fālala

Brush-sided leatherjacket, *Amanses scopas* (20 cm/8 in.)
pa'umalō, fālala

Yelloweye leatherjacket, *Cantherhines dumerilii* (35 cm/14 in.)
pa'umalō

Honeycomb leatherjacket, *Cantherhines pardalis* (20 cm/8 in.)
pa'umalō, fālala, aimeo

Fan-bellied leatherjacket, *Monocanthus chinensis* (38 cm/15 in.)

Beaked leatherjacket, *Oxymonacanthus longirostris* (9 cm/4 in.)

pa'umalō gutuumi

The beaked leatherjacket has a most wonderful appearance, with yellow polka dots spread perfectly over its blue-green body. Adults are small, usually travel in pairs, and are presumed to be monogamous. You might see these tiny jewels hovering close to *Acropora* coral, their food source. The branches of the coral also provide a safe place to hide for these hummingbirds of the fish world. Once you see one, this species will likely become one of your favorites.

Gill-blotch leatherjacket, *Pervagor janthinosoma* (14 cm/6 in.)

Black-headed leatherjacket, *Pervagor melanocephalus* (10 cm/4 in.)
paʻumalō, fālala

BOXFISHES (OSTRACIIDAE)

The boxfishes, also known as "trunkfishes," are peculiar in that they have a bony carapace (outer covering) of polygonal plates, with holes for their eyes. Boxfishes are triangular, quadrangular, pentagonal, or round in cross-section. Their tiny mouths are set low on the head and are surrounded by thick lips. Those with sharp, hornlike appendages on their heads are known as "cowfishes." Boxfishes have dorsal fins set far back on their bodies and lack pelvic fins entirely. They do not have scales, but rather fused, bony plates that encase their bodies in a boxlike shape. They are generally slow swimmers and resemble miniature helicopters buzzing along the reef.

Boxfishes are diurnal and feed on a variety of benthic animals. Some excrete a skin toxin called "ostracitoxin" when under stress. This toxin has been known to kill other fish in close association, such as in an aquarium, and to kill even the boxfish itself. Some species change coloration, especially during courtship dances.

Three species are known from the archipelago. The general Sāmoan name for boxfishes is **moamoa**.

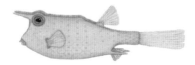

Longhorn cowfish, *Lactoria cornuta* (46 cm/18 in.)
moamoa ulutao or **moamoa uluto'i** (**uluto'i** means "protruding forehead")
Because cowfish move rather slowly, other species of fishes
will dash in and steal the cowfish's food; this species is more
likely to be observed in sandy or weedy habitats.

adult

juvenile

Yellow boxfish, *Ostracion cubicus* (45 cm/18 in.)
moamoa lega
The English translation of **lega** is "turmeric powder" or
"yolk of an egg," both wonderful descriptions of this darling
yellow cube-shaped fish (juvenile phase).

Spotted boxfish, *Ostracion meleagris* (18 cm/7 in.)
moamoa sama or **moamoa uli** (**uli** means "black")

PUFFERFISHES (TETRAODONTIDAE)

The pufferfishes, also known as "blowfishes," are so named because of their ability to inflate when frightened. Because puffers are relatively poor swimmers, inflation offers an alternate method of escape from predation. Water is drawn into a special area near the stomach, and the resultant sight is something quite rotund. Once the bulbous body is inflated, those species with spines—now positioned upright—are protected from predators. Those species without spines can balloon themselves into tight crevices for protection. Pufferfish have a becoming appearance and swim rather buoyantly across the reef. Most species are carnivorous. Puffers are known as "toadfishes" in Australia. The Tetraodontidae family is further classified into species belonging to the sharpnose puffer group, also known as "tobies."

Although considered a delicacy in some cultures, puffers produce a poison, tetraodontoxin, which is concentrated in certain organs of the fish. This is another way these slowpokes augment protection. While some human fatalities have occurred from eating puffers, toxicity varies with species, geographic location, and time of the year.

Fourteen species are known from the archipelago. The general Sāmoan name for puffers is **sue**.

Stars and stripes puffer, *Arothron hispidus* (48 cm/19 in.)
sue vaolo

Map puffer, *Arothron mappa* (60 cm/24 in.)

yellow phase

dark phase

Guineafowl puffer, *Arothron meleagris* (40 cm/16 in.)
sue puleuli (dark phase), **sue lega** (yellow phase)

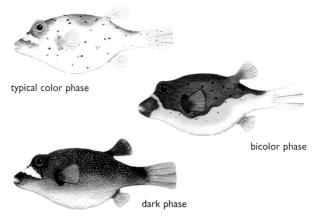

typical color phase

bicolor phase

dark phase

Blackspotted puffer, *Arothron nigropunctatus* (25 cm/10 in.)
sue uli (dark phase), **sue lega** (bicolor phase)

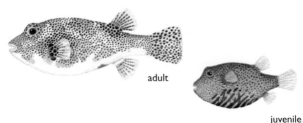

adult

juvenile

Star puffer, *Arothron stellatus* (90 cm/35 in.)
sue gatala, **sue va'a**

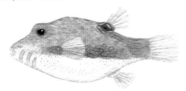

Bennett's toby, *Canthigaster bennetti* (10 cm/4 in.)
sue afa

Solander toby, *Canthigaster solandri* (8.5 cm/3.3 in.)
sue mimi
This species of sharpnose puffer is usually observed in pairs,
as are other species in the group.

Black-saddled toby, *Canthigaster valentini* (9 cm/4 in.)
sue mu

This species of pufferfish is impersonated by the mimic leatherjacket (*Paraluteres prionurus*). The mimic leatherjacket contains no skin toxin, but so closely resembles the poisonous black-saddled toby that predators avoid it too.

PORCUPINEFISHES (DIODONTIDAE)

Porcupinefishes, also known as "burrfishes," "balloonfishes," or "spiny puffers," are very similar to puffers, but have prominent spines on the head and body. The spines are highly modified scales that lie rather flat against the body unless the fish is inflated. These fish have a special internal water bladder that they can quickly fill. Once inflated, the fish enjoys protection against almost any enemy. It has been described as a puffed-up, spine-covered basketball with eyes and fins. Porcupinefishes have mouths that appear to be set in a precious grin and very large and beautiful eyes. Once you have seen them, you will become totally enamored with these fish. While most species are nocturnal, you can often find them peeking out from their daytime hiding places between rocks or under coral ledges.

Porcupinefishes have extremely strong jaws that they use to crush the hard shells of their prey, such as mollusks and urchins. These are among the slowest of fishes, appearing to buzz slowly along through the water. Although they are generally tolerant and passive, do not tease or catch one in order to inflate it or take a picture. Not only should you not mistreat the fish, but a wound from the spines can be painful, and a threatened porcupinefish will inflict a severe bite.

Predatory oceanic fishes feed on juvenile porcupinefishes at their own peril: large fish have been found dead with inflated porcupinefish stuck in their mouths.

Four species are known from the archipelago. One of them, *Diodon eydouxii*, is unlikely to be encountered, as it is pelagic during its entire life. The general Sāmoan names for porcupinefishes are **tauta** or **tautu**.

Freckled porcupinefish, *Diodon holocanthus* (29 cm/11 in.)

Porcupinefish, *Diodon hystrix* (71 cm/28 in.)
tauta, tautu

Black-blotched porcupinefish, *Diodon liturosus* (50 cm/20 in.)
tauta, tautu

MARINE INVERTEBRATES

Most of the animals of the sea are invertebrates (animals without a spinal column or backbone), in contrast to the vertebrates (animals with a backbone) we are so familiar with on land. They display an endless variety of forms, behavior, and ecology, and interact with each other and with fishes in every way imaginable. Some live in or on the bodies of other animals. Others build upon, burrow into, or tunnel through the reef, adding to its complexity. Still others remove parasites from fellow creatures or parasitize their neighbors.

Six categories of marine invertebrates are particularly abundant on reefs: (1) sponges, (2) polychaetes, (3) mollusks, (4) crustaceans, (5) echinoderms, and (6) cnidarians. Marine plants such as sea grasses or algae are also integral parts of the reef ecosystem and help make life possible for all other residents. A few common marine plants are illustrated below.

Turtle weed, *Chlorodesma* sp.

Coralline alga, *Halimeda* sp.

Sailor's eyeball (green alga), *Valonia ventricos*

143

SPONGES (PHYLUM PORIFERA)

Sponges are the oldest living group of multicellular organisms and have been on Earth for over half a billion years. They are simple animals that strain ocean water, filtering out oxygen and microscopic plankton, and thrive in all areas of the reef. Sponges may filter a volume of water equal to their own body volume every 20 minutes or so. They are entirely sessile, never visibly moving. Because of this, for a long time scientists thought sponges must be plants, so little did they resemble animals. Sponges come in every exquisite shade of the rainbow, and have both male and female organs. Snails, starfishes, nudibranchs, fishes, and some sea turtles eat sponges. Calcium carbonate spicules help support the sponge's body.

Encrusting sponge, *Leucetta* sp.

WORMS AND WORMLIKE CREATURES (PHYLA PLATYHELMINTHES AND ANNELIDA)

Coral reefs are teeming with various species of unassuming worms. Two of the phyla are the flatworms (Platyhelminthes) and polychaete worms (Annelida). Several other phyla of marine worms or wormlike creatures exist.

Flatworms

Flatworms are members of the phylum Platyhelminthes. They exhibit beautiful colors and might easily be confused with nudibranchs, although flatworms have a flattened body similar to a potato chip, while most nudibranchs are three-dimensional. Some flatworms acquire protection by mimicking the appearance of those nudibranchs that secrete harmful chemicals to discourage predators. Others actually are noxious. Those exhibiting the most glorious colors inform potential predators that the animal is toxic or distasteful. Others use camouflage as their survival tool. Most flatworms are tiny (< 8 cm). They move across the reef using microscopic bristles on their undersides, sliding along a trail of secreted mucus. They are also excellent swimmers if need be, and move through the water by undulating the edges of their bodies in rhythmic waves.

Flatworms can reproduce sexually or by regenerating an entire individual from a fragment of themselves.

Pseudoceros zebra

Pseudobiceros bedfordi

Polychaete Worms

Polychaete worms are members of phylum Annelida, class Polychaeta. Polychaete worms exhibit a variety of shapes and make their living in many different ways.

The feather duster and fan worms are common on most reefs. These worms live in secreted tubelike homes into which they disappear at the first hint of danger. Some tubes are visible, embedded in coral heads. As the coral grows, the worms secrete additional tubal material to keep pace with the growth of their coral host. Only the feathery feeding tentacles protrude from the tube, capturing tiny planktonic organisms floating by. Sexes are usually separate in polychaete worms, and eggs and sperm are simultaneously released by males and females.

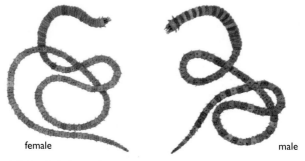

female

male

Palolo, *Palola siciliensis*
palolo
Females are blue-green; males are light brown.

Christmas tree worm, *Spirobranchus giganteus*
The Christmas tree worm is a type of feather duster worm (family Serpulidae) that comes in a variety of fanciful colors. While most polychaete worms are observed only at night, the delicate feeding gill tufts of the Christmas tree worm can often be seen during the day filtering plankton from the water. The slightest disturbance will make the worm withdraw into its tube.

In the Sāmoan Archipelago, one type of poly-chaete worm, the **palolo**, is an important delicacy and cultural symbol. The worms live in branching burrows in the limestone base of shallow tropical reefs. Although typically "active" at night, the worms apparently never leave their burrows.

The eggs and sperm of **palolo** concentrate in the tail segments of mature adults. Seven days after the full moon in October or November, when evening falls, the gamete-filled tail segments detach from the main body of the worm and float to the ocean surface above the reef to reproduce. In some areas, millions of the spaghetti-like **palolo** form a wriggling mass, permeating the water. Eventually the wriggling stops, and the tail segments dissolve into a slick of eggs and sperm.

This annual event is eagerly awaited by the Sāmoan people. Those **palolo** that are not swallowed on the spot by hungry fishermen are gathered for subsequent feasts. Ulas (flower necklaces) made from the flowers of the **mo'oso'oi** *(Cananga odorata)* tree, which blooms at the same time of year that **palolo** spawn, are often worn by fishermen at this time of year out of reverence for the event.

Interestingly, the actual time of emergence of **palolo** worms differs between the islands, earliest in the eastern end of the archipelago, and later as one goes west. They usually appear about 10 p.m. in Manu'a, about 1 a.m. on Tutuila, and not until about 4 a.m. in the western islands of Sāmoa. Although swarms occur in many other areas of the South Pacific, in Hawai'i, the **palolo** worm, though present, does not exhibit this mass spawning behavior.

MOLLUSKS (PHYLUM MOLLUSCA)

The mollusks (**figota**) are a diverse group of animals that includes the gastropods (snails or seashells, nudibranchs, sea hares), bivalves (clams, oysters, mussels, and scallops), and cephalopods (squid, cuttlefish, octopods, and *Nautilus* sp.). All have similar body plans, and all mollusks have a shell during some stage of development. The shell is secreted by a part of the body called the "mantle." Some mollusks lose their shells as adults (e.g., octopuses, land slugs, nudibranchs). Most mollusks feed using a radula, a flexible, tonguelike structure covered with tiny teeth, which they use to grate back and forth across a food source in order to remove small edible particles.

To date there are approximately 100,000 described species worldwide. Although a complete and accurate count of local mollusks has not been compiled, experts in this field of study estimate that over 2,000 species probably occur in the Sāmoan Archipelago. Giant clams, large marine snails (e.g., *Trochus* sp. and *Turbo* sp.), and octopuses are among the most important mollusks eaten locally.

Snails and Slugs: The Gastropods (Class Gastropoda)

Class Gastropoda is the largest in the phylum Mollusca and comprises approximately 67,000 species, 40,000 of which are marine. Two main groups (subclasses) of gastropods are discussed in this chapter: the Prosobranchia (snails or seashells) and the Opisthobranchia (nudibranchs, sea hares, bubble shells, and others). The third subclass is Pulmonata (land snails and slugs) (see Terrestrial Invertebrate chapter). Gastropod means "stomach-footed," and all gastropods have the same basic body structure.

Seashells (Sisi)

Seashells and snails are recognized by their single, hard, calcium carbonate shell. The calcium is extracted from seawater and food sources and deposited along the edge of the shell by an organ known as the "mantle." The resulting shell structure has coiled architecture. In addition, the mantle is responsible for maintaining the luster of the outer surface of the shell. Sea snails have a distinct head and eyes, sensory tentacles, and a strong muscular "foot." The large fleshy foot is used for locomotion. Some species have a hardened plate attached to the foot, called an "operculum." The operculum is used like a trap door to close off the shell from predators or keep the animal from drying out. Most species have a tonguelike structure called a "radula," covered with hard, tiny teethlike structures, which they use to scrape surfaces for their algal meals. Still others (the cone shells) have modified radula that they use as a venomous harpoon to inject poison into worms, other mollusks, or fishes.

Sea snails are not illustrated in this book, as the numerous species (approximately 2,000 from the Sāmoa Islands) deserve a book unto themselves.

Cone shells. Do not handle these animals, even if you are wearing gloves (the venomous harpoon of the cone shell has been known to pierce fabric). All cones are venomous; the fish-eating and mollusk-eating species are potentially most harmful to humans. One species, the geographic cone *(Conus geographicus)*, is responsible for several human fatalities each year. Other species can inflict dangerous and painful wounds. Holding a cone shell by its wide end is no guarantee you will not be impaled, as the harpoon (proboscis) organ is highly mobile and accurate.

Nudibranchs and Closely Related Species

Nudibranchs (**gaupapa sami**), or sea slugs, are one of the most beautiful and delicate mollusk groups. They are a subclass of Gastropoda, classified as opisthobranch mollusks. No one knows how many species of nudibranch there are, but scientists estimate between 1,000 and 3,000. Nudibranch means "naked gills," and refers to the feathery, gill-like structures apparent on the dorsal surface of many species. Certain species, known as the "phyllidiids," lack these prominent gills and have bumpy ridges on their backs instead. Others have fingerlike appendages or long flanges.

Unlike their relatives the snails, nudibranchs lack a protective outer shell. Some species do have a small, internal shell (e.g., *Haminoea* sp.). Nudibranchs are found in all reef habitats, from only a few inches of water to great depths. Most species range in size from less than 1 cm (.39 in.) to over 30 cm (1 ft.), and are usually cryptic in behavior. One type of sea slug, the sea hare (e.g., *Aplysia* sp.), can reach a length of over 38 cm (15 in.). Nudibranchs are usually either cryptic or well camouflaged, leading secretive lives under the coral rubble or hidden among coral branches. Others are nocturnal, emerging only under the cover of darkness. Many species sport tentacle-like structures called "rhinophores," which are chemosensory organs. Nudibranchs have both male and female reproductive organs within a single individual, and each nudibranch in a mating pair fertilizes the eggs of the other.

Of the species that are conspicuous and beautiful, the coloration is a warning that informs potential predators that their flesh is toxic or extremely distasteful. Once a fish inadvertently nibbles such a nudibranch, it is unlikely to make the same mistake again. The nudibranchs incorporate the toxins and stinging cells of their algal, coral, or sponge meals into their own tissue to use them in their own defense; the nudibranchs themselves are immune. Some nudibranchs also obtain their coloration from the food they eat. For example, a pink nudibranch that lives on pink coral has obtained its coloration from the very coral it is eating and is thereby camouflaged from predators.

There are five major orders of opisthobranchs, each describing an anatomical feature characteristic of members of the order.

1. Bubble shells (Cephalaspidea): Cephalaspideans are characterized by a head shield that is thought to have evolved as a burrowing aid. Typically, there are two lobes running along the length of the body, which can be used for swimming. Some cephalaspideans have a thin, bulbous, internal shell (e.g., *Haminoea* sp., *Cheliodonura* sp.).

2. Sacoglossans (Sacoglossa): Sacoglossans are characterized by a radula with only a single row of teeth. Most are herbivorous and feed by sucking out the contents of individual algal cells. Some species have functional chlorophyll in their own tissues, which gives them a brilliant green color (e.g., *Plakobranchus* sp., *Elysia* sp.). Chlorophyll is a pigment, typically found in plants, that converts sunlight into energy. The nudibranchs eat algae, thereby obtaining the green pigmentation.

3. Sea hares (Anaspidea): In contrast to the cephalaspideans, anaspideans are characterized by their lack of a head shield. They are commonly known as "sea hares." This common name refers to their long, earlike rhinophores, which can be 10–20 cm (4–8 in.) long. All anaspideans are herbivorous and feed on algae. Their fat bodies contain the remnant of an internal shell (e.g., *Aplysia* sp.).

4. Side-gilled slugs (Notaspidea): Notaspideans are named for the shape of their broad, flat, dorsal surface. Only some species have shells. All nudibranchs in this order are carnivorous (e.g., *Berthellina* sp., *Pleurobranchus* sp.).

5. Nudibranchs (Nudibranchia): As mentioned earlier, nudibranch means "naked gills." This order is the most high-

ly evolved and exhibits some of the most delightful combi-
nations of color and pattern seen in nature. All species have
a flat, footlike structure and anterior rhinophores. Other
structures, such as gills, may or may not be present. All nudibranchs are carnivorous (e.g., *Hexabranchus* sp.,
Chromodoris sp., *Phyllidia* sp., *Flabellina* sp.) and exploit a
unique food source: small, benthic filter-feeding organisms.

All nudibranchs are very delicate. Please leave them
where you find them: LOOK BUT DO NOT TOUCH.

Aplysia dactylomela
These are large close relatives of the nudibranchs commonly referred to as "sea hares." Many sea hares secrete purple
ink as a defense mechanism.

Berthellina citrina

Chromodoris aspersa

Chromodoris decora

Chromodoris fidelis

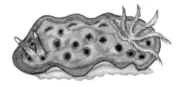

Chromodoris kuniei

Chromodoris lochi

Cheilidonura varians

Cyerce nigricans

Elysia ornata
Species of *Elysia* have functional chloroplasts in their tissue
that give them their green coloration.

Flabellina sp.

Flabellina rosea

Flabellina exoptata

Haminoea cymbalum
This group typically has an internal "bubble" shell.

Spanish dancer, *Hexabranchus sanguineus*
The Spanish dancer is one of the largest, most brilliantly colored nudibranchs. While crawling is its usual method of moving, it can also flare its winglike mantle and undulate its body in order to "dance" through the water. This nudibranch's egg case looks like a delicate fabric rose resting on the reef, but it is really a coiled ribbon of eggs. Adult Spanish dancers contain a chemical compound called "macrolide," which is an extremely efficient fish repellent. The chemical is extracted from their sponge meals and concentrates in the eggs.

Myomela sp.

156

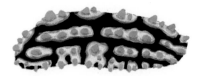

Phyllidia varicosa
This is thought to be the most common and widespread species of *Phyllidia*.

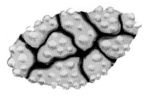

Phyllidia pustulosa

Phyllidiopsis striata

Plakobranchus ocellata

Phyllidia ocellata

Roboastra gracilis

Siphopteron tigrinum

Bivalves (Class Bivalvia)

In contrast to the single, coiled shell common in most gastropods, bivalve shells have two parts. Clams, oysters, scallops, and mussels are well-known bivalves. Because some bivalves are permanently attached to the substrate, they obtain their food by filtering tiny plants and animals from the water. Members of the giant clam family (*Tridacna* sp.) contain zooxanthellae (microscopic algae) in the tissue of their mantle, as do corals. The alga produces food for its host and imparts the beautiful varieties and patterns of colors we observe in individual clams. Members of the same species may be very differently colored, with shades of blue, purple, green, and brown.

The "giant" giant clam (*Tridacna gigas*) of Australia can reach a diameter of 1.3 m (4.3 ft.) and live up to 200 years. The giant clam species of the archipelago are quite a bit smaller. Unfortunately, these clams have been exploited in many areas, causing a significant decrease in their abundance in the archipelago and elsewhere in the South Pacific. One species, the bear paw clam (*Hippopus hippopus*), has been extirpated (become locally extinct) from our reefs. In some locations, other species hang on in precipitously low numbers. Because giant clams are highly valued as food, one species (*Tridacna derasa*) has been introduced to the Sāmoa Islands in hopes of establishing small aquaculture businesses. It is illegal to fish and keep giant clams less than 20 cm (7.9 in.) in the western islands of Sāmoa, or less than 15.5 cm (6 in.) in American Sāmoa. Further, in American Sāmoa it is illegal to sell any giant clams less than 18 cm (7 in.), and all clams offered for sale must be in whole condition, i.e., meat attached to shell.

Mollusks are an important and economically valuable fishery. Some species, such as the giant clam, are highly valued for food, while oysters are valued for natural and cultured pearls.

Fluted giant clam, *Tridacna squamosa* (40 cm/16 in.)
faisua

Small giant clam, *Tridacna maxima* (30 cm/12 in.), IUCN Red List: VU (imminently)
faisua

Southern giant clam, *Tridacna derasa* (50 cm/20 in.), IUCN Red List: VU
An attempt is being made to repatriate this species into Sāmoan waters.

Bear paw clam, *Hippopus hippopus* (40 cm/16 in.), IUCN Red List: VU (imminently)
It is thought this species is locally extirpated from the Sāmoas, though empty shells of this species are still occasionally observed lying on the bottom of some lagoons.

The Cephalopods (Class Cephalopoda)

Octopuses, squids, cuttlefishes, and nautiluses are all cephalopods. Cephalopods are evolutionarily advanced mollusks with the most advanced nervous system of all invertebrate animals. The word "cephalopod" means "head-foot" and refers to the animal's two dominant body parts, head and tentacles. This group of mollusks is generally adapted for swimming or rapid movement. Most cephalopods are known for their ability to eject a cloud of ink through a muscular siphon in order to confuse an attacker and veil their escape. Cephalopods also use the siphon as a source of jet propulsion by forcing water through its opening. Typically, the siphon can be directed to control the resultant direction of movement. The siphon of the nautilus, however, faces forward. The nautilus is, in essence, "swimming backward" and cannot see where it is going.

When disturbed or threatened, octopuses can rapidly change color to a mottled pattern, then back again. This may confuse a predator long enough to allow the octopus to escape. Octopuses are often so perfectly camouflaged as to be virtually invisible in their hiding places. In addition to changing their color, octopuses and cuttlefishes (a close relative of squids) can modify the texture of their bodies to match the surface of their surroundings. While some octopuses are active during the day, many are strictly nocturnal and, during the day, are usually seen hiding in a hole or under a coral head. They primarily eat crustaceans and other mollusks.

Squids are the fastest invertebrates, reaching speeds of up to 45 kph (28 mph). They are also masters of color and pattern change, especially during romantic encounters.

Nautiluses are the oldest, most primitive cephalopods, dating back over 400 million years. Only five species exist today, and all live in very deep water (60–750 m in depth). They undertake long-distance feeding trips toward the surface daily, using their multichambered shell to regulate their buoyancy. In some parts of the Pacific, nautiluses come to the reef at night to feed, returning to their deep-water homes before dawn.

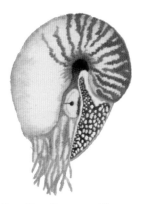

Chambered nautilus, *Nautilus pompilius*
Unfortunately, it is unlikely that you will encounter a chambered nautilus while snorkeling or diving, because it lives in very deep water.

Octopus, *Octopus cyanea*
feʻe

Bigfin reef squid, *Sepioteuthis lessoniana*
gūfeʻe

CRUSTACEANS (PHYLUM ARTHROPODA)

With approximately one million species worldwide, the arthropods are the largest phylum of living organisms on Earth. Members of the crustacean group (subphylum Crustacea) include such well-known animals as crabs, shrimp, and lobsters, as well as a host of other lesser-known types of organisms totaling perhaps 45,000 different species. The majority of crustaceans encountered on the reef are decapods, meaning "ten legs."

The phylum Arthropoda also includes terrestrial animals such as the insects, spiders, scorpions, millipedes, and centipedes (see Terrestrial Invertebrates chapter).

Crustaceans are abundant and active inhabitants of the reef, but they are also nocturnal and not frequently observed during the day. They have a hardened outer skeleton made of chitin and calcium carbonate. Some crustacean shells are thin; others are thick. Some are retained for life, while others are shed as the animal grows. When the animal periodically outgrows its shell or carapace, a portion of the shell's calcium is reabsorbed, and a soft, expandable new skin forms underneath. Once the old shell is shed or molted, the new skinlike shell is inflated with water to its full size. Because the new shell is initially soft, the now-vulnerable animal must find a safe place to hide until the new carapace hardens. While walking along the beach, you might notice what looks like a dead crab perched on a rock or piece of coral rubble. Look carefully; there is probably no telltale evidence that the animal was eaten (are there any large bites missing?). Rather, it is probably the shell of an animal that molted. Many crustacean eyes have clear, hard coverings that resemble light bulbs. The stalked eyes provide the animal with a very wide field of view, exceeding 180 degrees.

Even barnacles are members of the class Crustacea. Although they do not look anything like a typical crustacean as adults, and more closely resemble a bivalve mollusk, their relationship to other crustaceans is visible in the larval stage.

Hermit crabs ('u'a): Hermit crabs are not true crabs, but are more closely related to their crustacean brethren the lobsters. Some crabs seem to need more protection than just their own carapace. To protect their large, soft abdomens they seek safety by using the hard shells of others. Typically, they live in the shells of gastropod mollusks, although occasionally they use some other object, such as a broken bottle, as a shell. When hermit crabs molt, they simply find larger shells to hide in and do not have to wait for their new shells to harden before resuming their crab lives.

Coconut crabs: The coconut crab is the largest terrestrial invertebrate in the world. Individuals can measure up to 1 meter (39 inches) from leg tip to leg tip and weigh as much as 13 kg (29 lbs.) (Burggren and McMahon 1988). The coconut crab is actually a species of hermit crab, without the borrowed-shell home. Coconut crabs are omnivorous and will eat tropical fruits, fish, and other crabs. As their name implies, they also eat coconuts. How do they open them? The crabs husk the coconut with their strong chelipeds (claws), then pierce the weakest "eye" of the coconut. Once the coconut is husked, some crabs scrape the coconut out of this tiny (1 cm) opening, while others manage to chip off large pieces of nut. A coconut crab in Papua New Guinea was once observed husking a coconut, carrying it up a tree, then dropping it to the ground, where it cracked open. Whether this crab was particularly clever or just clumsy is not known. Coconut crabs are also known as "robber crabs" and are quite clever at stealing food from other crabs and even campsites. They have been known to wander off with anything they can carry. One was observed at the edge of a beach towing a whiskey bottle behind it.

Because of their delicious flavor, coconut crabs have been hunted almost to extinction on many Pacific islands. The IUCN *Red List of Threatened Animals* lists this species as possibly being considered vulnerable (facing a high risk of extinction in the medium-term future), but there is a need for more data before such a determination can be made. It is illegal to fish for and keep coconut crabs less than 7.5 cm (3 in.) across the widest portion of the back or any egg-bearing females, or to interfere with a crab releasing larvae.

Some island traditions call for careful harvesting of reef and land resources. For example, a particular species may be banned from being harvested periodically or during a certain time of year, thus ensuring the survival and sustainability of that resource.

In the Cook Islands, one type of traditional management called *rahui'i* is used specifically to conserve reef resources. The chief of a family or village can place a *rahui'i* on a particular area or species to prevent its exploitation at a particular time of year. A *rahui'i* is often indicated by a simple, physical sign, such as a coconut leaf tied around a tree on the way to the area that is off-limits. Coconut crabs are frequently put under a *rahui'i* because they are so vulnerable to overharvesting.

Coconut crab, *Birgus latro*, IUCN Red List: VU
ūū

Three-spot crab (also known as the "7-11" crab), *Carpilius maculatus*
kuku
This species, with its large red spots, is easy to identify. Its clawed legs are particularly large. It may be poisonous to eat. It is illegal to fish for and keep **kuku** less than 12 cm (4.7 in.) in the western islands of Sāmoa.

Rock crab (also commonly referred to as "Sally Lightfoot" crab), *Grapsus* sp.
ama'ama, **pa'a**
Rock crabs are extremely fast, capable of jumping from rock

to rock and quickly retreating at the first sign of danger. Their flattened body shape allows them to fit into tight crevices.

Ghost crab, *Ocypode ceratophthalmus*
aviʻiviʻi, paʻa
Ghost crabs are large, sand-colored beach crabs common throughout the Indo-Pacific. They are more terrestrial than marine, but can stand submersion in water for short periods. Their long legs propel them swiftly across the sand, enabling them to reach speeds of 1.8 m per second. They are also good diggers, and the large holes of their sandy burrows are visible along the beach. Large mounds of sand can be observed outside the burrow entrances of male crabs, while females tend to disperse the sand across the beach. Although primarily nocturnal, adult ghost crabs can be seen in the early morning hours in the intertidal zone. Juveniles are active throughout the day, but are so perfectly camouflaged that when they stop moving, they seem to disappear.

Green mangrove crab, *Scylla paramamosian*
paʻalimago
It is illegal to fish for and keep **paʻalimago** less than 15 cm

(5.9 in.) in the western islands of Sāmoa, or any egg-bearing females or crabs less than 15 cm (6 in.) in American Sāmoa.

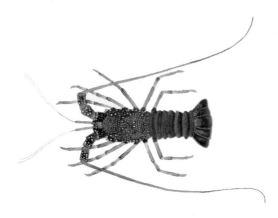

Golden rock lobster, *Panulirus pencillatus*
ula sami
It is illegal to fish for and keep any spiny lobster under 8 cm (3.2 in.; carapace length, measured from front to rear edge of carapace) in either Sāmoa.

Slipper lobster, *Parribacus antarticus* (also known as *Parrabacus caledonicus*)
papata
It is illegal to fish for and keep any slipper lobster under 15 cm (5.9 in., measured from front to rear edge of tail) in the

western islands of Sāmoa, or any egg-bearing slipper lobster in American Sāmoa.

Land hermit crab, *Coenobita perlatus*
While once common at Rose Atoll, this species may be in decline at this location, for reasons unknown at this time.

ECHINODERMS (PHYLUM ECHINODERMATA)

Sea urchins, sea stars (starfish), brittle stars, sea cucumbers, and feather stars all belong to the group collectively known as "echinoderms," which means "spiny skin." Echinoderms in general have radial symmetry, and the bodies are generally divided into five sections from which arms radiate off the central body. In most species, the body actually consists of five equal segments that each contain a set of internal organs. The mouth is centrally located on the bottom side (known as the oral side), and the anus is located on the top (aboral) side. Except for brittle stars and some starfish, locomotion is usually very slow. Echinoderms all have an internal, water-pressured vascular system that maintains the shape of their water-logged bodies and hydraulically controls their many tube feet. The tube feet of most echinoderms are equipped with suction disks at their ends. Echinoderm tube feet are multifunctional; they are used for respiration (breathing), moving across and holding onto the substrate, and passing food to their mouths. Urchins and sea stars also have small pincers, which they use to remove parasites that land on their own bodies.

The radial pattern of five sections is visibly different in the different classes of echinoderms. Sea urchins are usually spherical, but may be somewhat flattened. Sea stars are generally flat with five arms, although some species have many arms. The brittle stars have a small central disc from which radiate five serpentine and often prickly arms. Feather stars and sea lilies have a central calyx from which many pinnate arms radiate upward. The five-parted radial pattern is least visible in the bloated tubular bodies of sea cucumbers.

Sea Urchins (Class Echinoidea)

Sea urchins have a basic echinoderm body, including tube feet hiding under their spiny exterior. Urchins have fragile, thin-shelled, spherical bodies covered with spines. Urchins produce a calcareous "test," which is not quite a shell and not quite a skeleton. It is formed from calcium carbonate extracted from the seawater. The spines are connected to the test in a ball-and-socket configuration. Tiny muscles encircling these ball joints control the movement of the spines, assisting the tube feet in providing transportation across the reef and protecting against predators. Urchins are typically nocturnal, and primarily eat algae, though they will consume animal matter too. The urchin mouth is located in the center of its "oral" side (underside).

One unusual type of sea urchin found in the Sāmoa Islands is the sand dollar. It has a flattened oval shell with a starfish pattern imprinted in the upper side of the test. Although numerous, its spines and feet are tiny. Sand dollars live under the sand.

Echinothrix diadema
vaga
The spines of this urchin are black and may have a blue-green, iridescent sheen when viewed in sunlight. These urchins are commonly seen during the day hiding in coral crevices or under ledges. At night, they come out of their hiding places to patrol the lagoon for food. Do not handle these or any other urchins; wounds from the spines can cause weeks of discomfort.

Slate pencil urchin, *Heterocentrotus mammillatus*
satula
This urchin is known for its fat, reddish spines, which often wash up on the beach. The spines are also used in some traditional handicrafts.

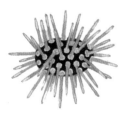

Mathae's burrowing sea urchin, *Echinometra mathaei*
tuitui ("poke, poke")
Color is variable: gray, red, black, or dark green.

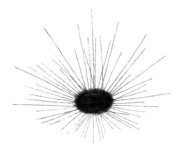

Diadema setosum

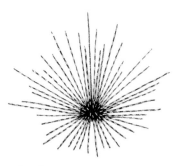

Diadema savignyi
vaga
This urchin's spines are broadly banded in black and white.

Sea Stars (Class Asteroidea)

Sea stars ('**aveau**, "arms that radiate") have a central, disc-shaped body with radiating arms. They are commonly referred to as starfish, and range in size from tiny (2 cm) to large (1 m). Typically, there are five arms, but some species can have up to forty. Sea stars move easily with the assistance of their tube feet, which extend from grooves radiating from the mouth on their oral surface. The aboral (opposite to or away from the mouth) surface is usually covered with bumps, knobs, spines, or sieves. Sea stars are carnivorous and eat almost anything. Some will even turn their stomachs inside out, envelop their prey, and when the meal is finished, retract their stomachs again. Sea stars reproduce in a variety of ways. Most species reproduce sexually: individual males and females release sperm and eggs into the sea, where the larvae then develop. Some species divide the disc into halves in a process known as "fission," with each half regenerating the missing body parts. Others simply cast off arms that can then regenerate a new disc and an entire complement of arms.

Like geckos and other lizards, some sea stars can cast off an arm to escape a predator, then grow it back later, in a process known as "autotomy." Most sea stars can be handled, except the crown-of-thorns, whose spines inflict painful wounds (see below).

Crown-of-thorns: The crown-of-thorns starfish has between 7 and 21 radiating arms, depending on the individual, and can live up to eight years. Both juveniles and adults feed on live coral polyps. Periodic population explosions of this animal lead to huge amounts of coral being devoured, and leave entire reefs destroyed. Such an outbreak occurred on the reefs in American Sāmoa on the island of Tutuila and in the western islands of Sāmoa during the late 1970s. The reefs of the Manuʻa Islands and Rose Atoll and Swains Island were spared the infestation. Crown-of-thorns are a natural part of an island's reef ecosystem and in low numbers do not appear to do significant damage. Sometimes small sentinel crabs live among the branches of corals. When a crown-of-thorns sea star approaches the coral, the crab emerges and pinches the sea star until it changes direction and moves off.

The wounds from their spines or even a single spine can produce pain that lasts for weeks. A widely known folk remedy, **e fofo e le alamea le alamea** ("the one who caused the injury is the one to make it right") recommends that if you are wounded, you should place the mouth of the very same crown-of-thorns onto the wound to remove its offense. For those for whom this remedy fails, the reply is always that you obviously must not have used the same offending animal.

The reasons underlying the periodic infestations by the crown-of-thorns are still largely a mystery. However, a single starfish can produce 65 million eggs, and if some environmental condition enhanced the survival of the larvae, an infestation could result. Some scientists speculate that infestations are a result of eutrophication from agriculture or other anthropogenic (human-caused) runoff onto the reefs. Eutrophication occurs when the water becomes unnaturally nutrient-rich in chemical compounds that should not be there. This condition leads to an abnormally increased number of phytoplankton, upon which the larvae feed.

Other scientists suggest that these infestations are entirely natural, and not a result of human activities negatively affecting the reef ecosystem.

What nutrients or other nonpoint source pollution could be running onto Sāmoa Islands reefs? Where is it coming from? How can it be stopped?

Crown-of-thorns, *Acanthaster planci*
alamea

Cushion star, *Culcita novaeguineae*
This wonderful pincushion sea star comes in a variety of
shades. Its bloated, pentagonal body can be found resting in
sandy areas of lagoon habitats. The radial pattern is more
apparent in the juvenile, which is flattened in shape.

Fromia monilis
'aveau

Linckia laevigata
'aveau
This large, periwinkle-blue sea star is often found on the reef flat, its arms and body contorted and conforming in shape to any surface it clings to. Also illustrated is a *Linkia laevigata* "comet," a regeneration of the entire animal from a fragment of itself.

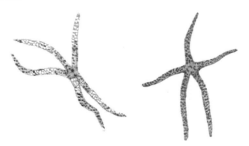

Linckia multiflora
'aveau

Brittle Stars (Class Ophiuroidea)

Brittle stars (**'aveau ma'ale'ale**) are closely related to starfish but differ in several respects. Brittle stars have only one set of organs in their central disc, and their typical set of five wiggly, flexible arms is made of many segments. In many species, the arms are easily broken off, hence their name. These same arms maneuver the brittle stars quickly across the substrate to the nearest hiding place. Their tube feet lack suckers and instead produce a slimy mucus to trap food.

Ophiolepis superba

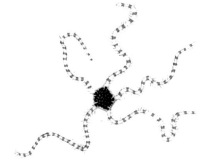

Ophioarthrum sp.

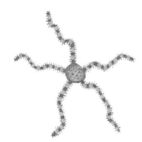

Ophiomyxa sp.

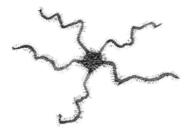

Ophiomastix sp.

Ophiarachnella gorgonia

Some scientists speculate that seemingly cata-strophic events that destroy entire reef ecosystems, such as starfish outbreaks or hurricanes, actually lead to the diversity of corals found on coral reefs. This may be true. In general, ecosystems rebound naturally from many earthly events as long as such events are not unnaturally frequent or a result of negative human activities. Although reefs can recover from some human-caused damage over time, if too many or too frequent human-induced destructive events take place in an ecosystem, long-term, and probably permanent negative con-sequences will prevail over time.

Sea Cucumbers (Class Holothuroidea)

Unlike their sea urchin and sea star relatives, holothurians do not have outward radial symmetry. Nevertheless, their relationship to those echinoderms is evident in their five rows of tube feet. They are sluggish animals with bloated tubular bodies that appear to be lying motionless on the substrate, hence their common name "sea cucumber."

Worldwide there are over 900 species. They range in size from 2.5 cm (1 in.) to 1 m (39 in.). Many sea cucumbers have a rubbery skin and tube feet on their ventral sides that help them cling to and move along the substrate. Unlike their echinoderm relatives, sea cucumbers have a mouth at one end and an anus at the other, although at first glance it is hard to tell one end from the other. A ring of tentacles surrounds the mouth, continuously ingesting sand, detritus, and organic material, and expelling a telltale trail of tube-shaped sand behind. Sea cucumbers "breathe" through their anus.

Some sea cucumbers brood their eggs on the surface of their bodies, while other species release embryos that develop in the water. Sea cucumbers have the ability to spurt internal, spaghetti-like threads called "tubules of Cuvier" from their anuses to discourage predators who try to molest them. The sticky threads probably help prevent further assault. As you move across the reef, it is not unusual to see these tubules floating by or snagged on a piece of coral. While not too dangerous, the tubules are toxic and may cause a slight rash. After an expulsion, new tubules will subsequently be regenerated by the animal for its next encounter.

Some sea cucumbers, such as leopard sea cucumbers, host symbiotic pearlfish who live inside their body cavities. At night, the pearlfish leaves the safety of its "home" in the anus of the sea cucumber to feed on the reef. If threatened, the fish will head to the rear end of the sea cucumber and tickle its anus with its slender tail. When the sea cucumber relaxes, the pearlfish slips back inside, tail first. It is thought that as many as 15 pearlfish can reside in a single sea cucumber. In addition to having a safe place to live, some species of pearlfish may actually parasitize their host for nutrients.

Some species known as "bêche-de-mer" are used in Asia as a base for soups. When cooked, their rubbery skin becomes quite gelatinous. Holothurians are pollution indicators, being particularly abundant where there are substantial unnaturally occurring nutrients such as untreated waste and other nonpoint source pollutants entering the water.

Actinopyga mauritiana

Leopard sea cucumber, *Bohadschia argus*
fugafuga

Holothuria atra
loli
This species is uniformly black or dark brown, with a smooth body frequently covered with sand grains.

Holothuria hilla

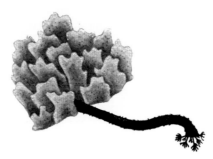

Holothuria leucospilota
This black, tubelike species has very visible tentacles surrounding the mouth. It lives on shallow reefs and feeds by extending its mouth and body over the sand while its posterior end remains hidden in coral or rocky substrate.

Stichopus chloronotus
amu'u, loli
Look closely to see the small orange tips on the many elongated knobs (papillae) protruding in two rows along the dorsal surface of the animal.

Synapta maculata
peva ("someone who does not have a skeleton")
This species is a snakelike, harmless tube of water. Its skin has hundreds of small hooks, much like Velcro®. Feathery tentacles surround the mouth, constantly consuming water and nutrients. Its accordion-like form enables it to move rapidly by inflating, then retracting its body.

Holothuria edulis
This species is observed in great numbers at Rose Atoll.

Crinoids: Feather Stars and Sea Lilies (Class Crinoidea)
Crinoids, a type of echinoderm common on some areas of the coral reef, are usually observed at night. At first glance, they resemble plants, not animals. They are filter feeders, using their arms to capture food particles and plankton floating by. The number of arms ranges from 5 to 200. Most hold on to the reef substrate with smaller, modified

arms called "cirri." Some species even walk, albeit slowly, to exposed food-rich edges of the reef at night for better filter-feeding opportunities.

Comanthus parvicirrus
This crinoid is commonly dark brown with pale-tipped, blue-gray pinnules; in some areas of its range, however, it may vary in coloration.

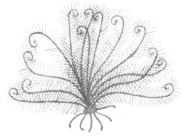

Comaster multifidus
This species is also variable in color.

PHYLUM CNIDARIA (FORMERLY COELENTERATA)

The creatures belonging to this phylum are the stingers of the underwater world, akin to transparent, watery wasps. There are over 9,000 species worldwide, all armed with stinging apparatuses called "nematocysts." There are three general types of nematocysts: stinging, adhesive, and entangling. Though not all nematocysts are venomous, in general, a nematocyst is a microscopic cell that contains a small bag of venom and a coiled structure with a harpoon-shaped stinger that is effortlessly thrust into the skin of its victim. Some of these stings can be extremely painful or dangerous, depending on the species. At the same time, this phylum contains some of the most beautiful and delicate of animals, no matter how noxious.

The cnidarians are further divided into three general classes: Anthozoa (stony corals, sea anemones, octocorals), Hydrozoa (fire corals, Portuguese man-of-wars), Scyphozoa (jellyfish, sea wasps).

Class Anthozoa

Anthozoans are the most abundant group of cnidarians. This class comprises the typical hard (scleractinian) corals; black corals; sea anemones; octocorals (soft corals), including the widely photographed gorgonians (sea fans); and sea pens.

This class is further divided based on the number of tentacles on the polyp: most have six, or multiples of six. The polyps of octocorals, as their name implies, always have eight tentacles. Information on the ecology and biology of corals can be found in the Coral Reefs chapter.

Hard corals (Scleractinia): Hard corals (also called stony corals) are the builders of coral reefs. Coral reef scleractinians are a diverse and abundant component of coral reefs and are what we typically think of when someone mentions coral. There are approximately 500–600 species of hard corals in the Indo-Pacific region, compared with only about 50 or 60 species in the Caribbean. Over 200 species have been observed on the reefs of the archipelago.

Blue coral: Blue coral is a type of cnidarian called an "octocoral" and is very common at the Va'ota Marine

Reserve on the island of Ofu, and in the Ofu unit of the National Park of American Sāmoa. It has a hard skeletal structure much like the true, stony (scleractinian) corals, but it is more closely related to the soft corals, sea fans, and sea pens. The blue coloration comes from iron salts extracted from its surroundings, but only the skeleton is blue. When alive on the reef, the living tissue is brownish. You will, nevertheless, see pieces of its blue limestone skeleton washed up on the beach.

Anemones: The gently rounded, beautifully flowing tentacles of the anemone (**matamalu,** "soft eyes," **lumane**) provide a safe place to hide for clownfish. The clownfish develops a special mucous coating on its body that renders it immune to the stinging tentacles of its anemone host. This is an example of a "commensal" relationship (see Coral Reef and Nearshore Fishes chapter), in which two species live in close association; one species benefits from the association (the clownfish) without harming the other species (the anemone). When in danger, some species of anemone can detach themselves from the substrate and somersault out of the way of danger.

Acropora sp.
This genus commonly occurs in antler (illustrated) and shelf forms.

Fungia fungites
This species looks like the inverted cap of a mushroom. Actually, this free-living coral comprises a single, very large polyp. Because it is not fastened to the substrate, it may be moved by wave action.

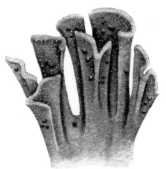

Blue coral, *Heliopora coerulea*
Blue corals are protected by law under the Convention on International Trade in Endangered Species (CITES). The skeleton, not the live coral, is blue. Look for skeletal pieces of dead blue coral washed up on the beach.

Montipora sp.

Brain coral, *Platygyra* sp.
Colonies may be spherical (illustrated) or flat.

Pocillopora sp.

Porites sp.
This genus forms huge boulders and microatolls.

Mushroom leather coral (soft coral), *Sarcophyton* sp.

Turbinaria sp.

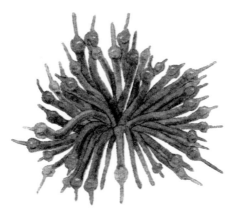

Bulb-tentacled sea anemone, *Entacmaea quadricolor*

Leathery sea anemone, *Heteractis crispa*

Class Hydrozoa

The hydrozoans come in a variety of forms, almost perfectly mimicking the body forms of hard corals. Hydrozoans occur as polyps (sedentary) or medusae (free swimming and jellyfish-like).

Fire corals are common sedentary hydrozoans that are easily mistaken for the hard (scleractinian) corals that are so abundant on the reef. Fire corals secrete a hard limestone skeleton, organize themselves in a colony, and exhibit a variety of shapes similar to the scleractinian corals. While you should not touch any of the corals or coral-like organisms, it is especially important that you not even brush against this creature. Its sting is excruciating, as if a burning object were being held against the skin. Depending on the severity of contact, the pain can last for several days.

Fire coral *Millipora* sp.
The sting received from brushing against a fire coral can last for days.

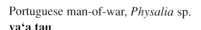

Portuguese man-of-war, *Physalia* sp.
va'a tau
Interestingly, the Portuguese man-of-war is actually a colony of hydrozoans, although its outward appearance is that of a jellyfish (class Scyphoza). These unusual hydroids have a special inflated sac that permits them to float across the sea. Although the floating structure measures only about 5 cm (2 in.), the tentacles, which are only about 30 cm (1 ft.) long when contracted, can extend up to 10 m (33 ft.). These tentacles trail from the floating portion of the animal and are often difficult to see against the surface of the sea. Their nematocysts can inflict a potentially fatal sting.

Class Scyphozoa

These are the animals that come to mind when someone says the word "jellyfish" (**'alu'alu,** "go away, go away"). These spectacular creatures are some of the most beautiful in the ocean. While they are eaten by some species of sea turtle, they seem to have few other predators.

Cubozoans are also members of class Scyphozoa, order Cubomedusae. These mighty stingers are the box jellyfish and sea wasps. Consistent with their name, they are cube-shaped, with four sets of tentacles hanging from each corner. Although they have not been observed in the waters surrounding the main islands, box jellyfish have been observed at Rose Atoll. Although also not known in Sāmoan waters, one type of sea wasp (*Chironex* sp.) from Australia may be the most venomous marine animal known, with a sting that can be fatal to humans.

There is a fabled animal in the archipelago that may actually be a jellyfish or cubozoan. It is known locally as **uila sami** ("ocean lightning"). Sāmoans say it can cause blindness if it stings you on the face, and those who have been stung (apparently only at night) have experienced excruciating debilitation. No physical descriptions exist for this denizen of the deep.

Over 90% of the 550 or so flowering plant species found in undisturbed (primary) rain forest and coastal habitats in the archipelago are indigenous, or native. There are an additional 230 or so species of ferns and fern allies (nonflowering plants) and over 100 species of orchids in the rain forest habitats of the archipelago. Further, approximately 30% of all the plant species are endemic to the archipelago (i.e., found nowhere else), with some endemic to a single island. Even a single plant genus, *Sarcopygme*, is endemic. An additional 250 species of accidentally or intentionally introduced plants that have become "naturalized" are referred to as weeds.

Most of the high volcanic islands of the archipelago were probably covered with dense rain forest before the arrival of Polynesians. The only major disturbances to affect the land were occasional hurricanes or landslides. When Polynesians settled the islands, they cleared much of the coastal and lowland rain forest for villages and plantations. Over the years, more primary forest has been eliminated. Today, the trend continues unabated as the population grows.

Rain forest (**le vaomatua**, **le vaomāoa**) is the natural vegetation of most of the archipelago. Botanists have separated the rain forest into 17 distinct categories, or types. Each type is based on the dominant tree or plant species in a particular location. For example, **tava**, **maota**, and **mamalava** trees dominate lowland rain forest; **vivao** dominates cloud forest. For simplicity's sake, the rain forest can be divided into four broad categories: coastal, lowland, montane. and cloud. As its name implies, coastal rain forest (**le vaomatua i le matafaga**) occupies areas along the coast. Lowland rain forest (**le vaomatua laugatasi maualalo**) once covered the low-lying lands and foothills. Today, lowland rain forest throughout the archipelago is severely threatened by land use conversion and, specifically, by logging in the western islands of Sāmoa. Montane rain forest (**le vaomatua i le mauga**) generally occupies the slopes of the mountains at higher elevations, up to 670 m or so. While trees may grow to great heights across mountain slopes and plateaus, along the mountain ridges (**le vaomatua tuasivi**) trees grow to only half that height. Cloud forest (**le vaomatua aua'oa, mauga pua'oa**) occurs at elevations exceeding approximately 670 m. It should be noted that there is usually no clear delineation between forest types, and species composition changes gradually with elevation and amount of rainfall.

More vines (**fue**) and climbing plants are found in the rain forest than in any other type of forest. Most large trees are adorned with vine "necklaces." An abundance of lianas (woody vines) winds and twirls through the forest trees, often providing a suitable surface for epiphytes such as mosses, ferns, and orchids. High above most other trees, strangler figs (*Ficus* sp., **aōa**) dangle aerial roots that wrap, hug, and coil around the trunks of other trees. Strangler figs (banyans) are actually a huge type of epiphytic tree that frequently begins life as a seed deposited within bird droppings on the surface of another tree. If conditions for growth are just right, the strangler seedling sends its roots down its support tree, hugging the bark. The roots eventually fuse together until they enclose and smother their support tree. As the host tree tries to grow, it is crushed to death. As the limbs of the tree grow outward, the banyan continues to send down aerial roots to support the ever-widening, expanding tree.

Towering trees with buttressed trunks are another extraordinary feature of tropical rain forests. Buttressed trunks provide the trees with secure footing even on rock surfaces where the layer of soil is thin. They also provide additional support against the fierce winds of tropical storms, helping to keep the trees upright.

Where sunlight is abundant, palms and tree ferns (**oliolī**) reach elegantly up to the sky like giant green, willowy umbrellas. The similar-looking giant ground ferns (*Cyathea* sp., **laugasesē**) also love light and moisture and can be seen growing in these same open areas. In disturbed areas of the rain forest, other ferns, vines, shrubs, and weeds quickly take root.

An epiphyte is a plant that uses another plant, typically a tree or liana, for its physical support. It does not draw nourishment from its host, and is therefore not a parasite. Staghorn fern (Platycerium sp.) is a common example of an epiphyte found in Sāmoan rain forests. Epiphytes are also known as "air plants."

Hurricanes Ofa (1990) and Val (1991) had a devastating impact on the forest, particularly in the western islands of Sāmoa. Many large trees were knocked down, and up to 90% of the trees were stripped of their foliage and fruit.

Cloud Forests

The cloud forests (**le vaomatua auaʻoa, mauga puaʻoa**) of the archipelago typically occur on summits of mountains at elevations exceeding 670 m (2,198 ft.). Mt. Lata on the island of Taʻu in American Sāmoa has such a cloud forest. The Silisili cloud forest in the center of Savaiʻi is the only cloud forest in the western islands of Sāmoa, but covers 80 sq. km (31 sq. mi.). This forest is the least-disturbed ecosystem in the western islands of Sāmoa because of its relative inaccessibility. The elevation at which cloud forest begins is somewhat arbitrary. For instance, a small portion of the forest above 550 m (1,804 ft.) on the island of Olosega is considered cloud forest.

The cloud forest is a misty world of dampness. Rainfall may exceed 1,015 cm (400 in.) annually. There is no dry season, and humidity is high year-round. The extremely wet soil, hollows, and cavities nurture the luxuriant growth of ferns and orchids. Epiphytes, though common in the rain forest at lower elevations, occur in amazing profusion in the cloud forest. Almost every trunk and branch is covered by mosses and liverworts, making the trees appear several times wider than their actual girth. Though tree ferns usually require open areas with a considerable amount of sunlight, they are common in the cloud forest on Ta'u.

Primary versus Secondary Rain Forests

Primary rain forest (**le vaomatua e le'i fa'ato'aina, o le ulua'i vaomatua**) has never been cut, degraded, or altered by anything other than natural causes (e.g., hurricanes, lava flows). Conversely, secondary rain forest (**vaomatua ua faatoaina**) habitat is the regrowth of forest after agriculture or other disturbed land has been abandoned. Secondary rain forest comprises an extremely large proportion of forest vegetation on all Sāmoan islands.

Secondary forest consists of all stages of regrowth. If left alone long enough (hundreds and hundreds of years), secondary forest may revert to forest virtually indistinguishable from primary forest. Nevertheless, this primary-like forest will regenerate only if seeds from the original species of plants are still available to propagate the forest.

Because of population pressures such as housing and agriculture, almost all of the Tafuna plain on Tutuila has been cleared of its native coastal and lowland rain forest. Similarly, almost all native lowland forest on the western islands of Sāmoa is gone.

Mangrove Forests

Mangroves inhabit not rain forests, but tidal forests,

growing near or between low and high tidal levels. Wetlands that are affected by tides are called "estuarine wetlands." Mangrove forests (**togā togo o i le pala**) or swamps (**pala**) can be found occupying the muddy, salty shores of sheltered lagoons along the coast. Mangroves are halophytic (salt loving) and are found only in brackish, or saltwater, conditions. The mangrove forests of Sāmoa are dominated by two species, the Oriental mangrove (*Bruguiera gymnorrhiza*, **togo**) and red mangrove (*Rhizophora mangle*, **togā, togo mumu**).

All mangroves are shallow rooted, and their stiltlike roots provide stable footing. The soils in which mangroves thrive are saturated with water, so, to prevent smothering (i.e., "drowning"), mangroves have characteristic aerial roots that "breathe." In addition, other roots absorb the nutrients mangroves need to grow.

Many streams are slowed down at the ocean's edge by mangrove swamps. As the water slows, the sediment drops out of the water and builds up the land. The mangrove forests are, in effect, one of nature's filters. They sift soil and other debris from the streams, thereby helping to ensure that clean water enters the reef. The sediments and nutrients are absorbed by the sea grasses and other vegetation in the swamp. The nutrient-rich habitat of mangrove swamps is critically important as nursery habitat for many kinds of fishes. Some saltwater fish species spend the early part of their lives in the mangroves, within the safety of quiet lagoon waters, before moving off to live on the coral reef. Mangrove swamps are also important for subsistence use, and some provide an important crab fishery as well.

Many of the historical mangrove forests of the islands of Sāmoa have been destroyed and filled in to create additional land (e.g., much of the village of Pago Pago in American Sāmoa, and Apia on the western island of 'Upolu). Other swamps are used as dumping grounds. Once mangrove swamps have been cleared and filled, they can no longer perform the important function of filtering the water before it flows onto the reef, nor do they provide habitat for crabs or fishes. Unfortunately, the filling of mangrove forests is still occurring at an alarming rate.

MAMMALS

In general, most South Pacific islands located far from continents have no or only a small number of native land mammals. Land mammals have few natural ways to disperse across wide expanses of water, and island biodiversity (the number of different species, both animal and plant, that are present) decreases as an island's distance from a continental land mass increases. Also, the size of the island plays an important role in its biodiversity; that is, the larger the island, the greater the number of species that can live there and flourish. Climate, an island's age, and a host of other factors further influence island biodiversity (see Carlquist 1974).

BATS

An animal that flies can cross water and colonize an island more easily than one that can't. Consequently, the only terrestrial mammals native to the archipelago are bats (**pe'a**). Bats belong to a group of animals known as Chiroptera ("hand-wing"). They are appropriately named, since their wings are actually elongated fingers with a membrane of skin and muscle between. Bats are covered with a soft coat of fur, and nurse their young with milk.

Bats are divided into two general groups: megabats and microbats. Megabats have large, sensitive eyes that provide excellent vision, even in nearly complete darkness. Microbats are generally smaller and use ultrasound, known as "echolocation" (emitting clicks and other sounds and listening for the echo), in addition to vision to "see" in the dark. Others have large, sensitive eyes that provide perfect vision in complete darkness.

Three species of bats occur in the archipelago. The Sāmoan flying fox and white-collared flying fox are megabats, with wingspans about 1 meter (3 ft.). They eat mostly fruit and nectar. They have large, light-sensitive eyes. The sheath-tailed bat, a tiny microbat with a wingspan of only 20 cm (8 in.), eats insects and uses echolocation. It is nearing extinction in American Sāmoa, and a small number are clinging to existence in the western islands of Sāmoa. Unfortunately, the sheath-tailed bat seems to be endangered throughout most of its range, including the Mariana Islands.

The archipelago's bats are found only on its high volcanic islands, because food and habitat requirements preclude them from living on Rose Atoll or Swains Island. All bats are extremely beneficial to the environment. Insect-eating bats consume thousands and thousands of pest insects, such as mosquitoes, each night. Fruit bats play an important ecological role as pollinators and seed dispersers in island ecosystems. Flying fox fruit bats feed solely on pollen, fruit, and leaves and play essential roles in dispersing seeds and pollinating the flowers of trees. They are the only known pollinators of the kapok (**vavae**) tree in the archipelago. They also are the most important reforesters of clearings. By scattering many thousands of seeds, fruit bats help "plant" new trees and help maintain the abundance and distribution of species in our rain forests.

The flying foxes and other bat species reproduce very slowly, typically giving birth to just one young a year. The white-collared flying fox breeds year-round, while the Sāmoan flying fox is observed most frequently with young during May and June, and rarely in December or January. Although weaned at three to four months of age, young flying fox fruit bats may remain dependent on their mothers for several more months, sometimes up to a year.

Flying foxes were once relatively common. In the early 1980s, however, hunters throughout the archipelago shot thousands of flying foxes to be sold to Guam, where they are considered a tasty delicacy. In a single three-year period, 18,000 dead flying foxes were shipped from the Sāmoa Islands to Guam. Bat numbers were further reduced as a result of hurricanes in 1987, 1990, and 1991, when trees were stripped of leaves and fruit. While many may have starved from a lack of fruit on trees, even more were killed by people, cats, and dogs when they were forced to uncharacteristically crawl on the ground near villages in search of food. Local law prohibits the hunting of flying foxes in American Sāmoa, but some are still hunted illegally. The number killed each year is unknown. Because the Sāmoan flying fox is quite active both day and night, it may fall victim to the hunter's gun more often. In addition, the Sāmoan flying fox tends to be solitary. The white-collared flying fox is primarily nocturnal; during the day it can be found roosting in colonies of hundreds of individuals. It is therefore possible to shoot numerous bats of this species in a short period of time. Commercial trade in both species of flying fox was banned in 1989, under CITES, but the independent nation of Sāmoa has yet to sign the treaty.

As of the year 2000, American Sāmoa Department of Marine and Wildlife Resource scientists estimate that there are fewer than 5,500 white-collared flying foxes in American Sāmoa. The Sāmoan flying fox population is precariously lower, estimated at only 900 bats or fewer. American Sāmoa currently prohibits any hunting of either flying fox species.

There are no current estimates for the number of flying foxes in the western islands of Sāmoa. Nevertheless, the government, in its State of the Environment Report, lists the Sāmoan flying fox as being at high risk of extinction. While the commercial export of flying foxes is prohibited in the western islands of Sāmoa, hunting is still allowed for local consumption. These flying foxes need to be surveyed periodically to see that they are not overhunted.

Forest and habitat quality is important to flying fox recovery. The National Park of American Sāmoa was established in part to provide protected habitat for flying foxes. While today it is unlikely to see a flying fox soaring above the trees in downtown Pago Pago, on a good day in the park you might be able to see a dozen. The best times and places to see flying foxes are early morning or late afternoon from roads and high passes in the National Park on Tutuila.

Sāmoan flying fox, *Pteropus samoensis,* IUCN Red List: VU

pe'a vao

The Sāmoan flying fox is quite rare and found only in the Sāmoan and Fijian archipelagos. While it is relatively active at night, this species is particularly unusual because it is also frequently active during the day. When not soaring over forests and valleys looking for food, it is likely to be roosting alone in forest trees. The Sāmoan flying fox typically has dark-brown body fur and wings that are broad relative to their length. Head and face coloration are variable. The top of the head and face of some individuals is a silvery, light gray; of others, russet. The shoulders and neck of males are dark auburn. This species has a shorter snout than the white-collared flying fox. In flight, its silhouette is rather triangular (compare with white-collared flying fox, below), and it has slow, shallow wingbeats. Males are slightly larger than females. The Sāmoan flying fox is listed internationally by CITES as an endangered species. CITES legally prohibits member nations from engaging in international commerce in species on the list.

White-collared flying fox, *Pteropus tonganus*
peʻa fanua, peʻa fai, taulaga
The white-collared flying fox is the most abundant species of
bat in the archipelago. It is also known as the Tongan flying
fox, and is widely distributed throughout the Pacific Islands.
Unlike the Sāmoan flying fox, this species of bat roosts
together in tree colonies of hundreds or thousands of individuals. In addition, like all other species of bat in the world, it
is mostly nocturnal. It leaves its roost beginning about an
hour before nightfall in search of food, often foraging across
large distances. It is one of only a few species of flying fox
that gives birth at any time of year. This species has a much
darker face than the Sāmoan flying fox, a longer snout, and a
distinctive white or yellow mantle that wraps from ear to ear
across the shoulders. Its wings are narrower for their length
than those of the Sāmoan flying fox, and the silhouette in
flight is more cross-shaped than triangular. The white-collared flying fox also exhibits a more rapid, deeper wingbeat.

Sheath-tailed bat, *Emballonura
semicaudata*, IUCN Red List: EN
tagiti, peʻapeʻa, peʻapeʻa vai

This species is a tiny, nocturnal, insect-eating bat; it does not eat fruit. Sheath-tailed bats usually roost colonially in deep, protected caves or lava tubes during the day, emerging at dusk to hunt for insect prey. Because these bats eat enormous numbers of small insects, such as mosquitoes, beetles, and moths, they are very beneficial to humans and agricultural crops.

A survey of their population in American Sāmoa in 1976 reported considerable numbers of bats, mostly living in caves along the shoreline, near the village of Afono on the island of Tutuila. During the mid-1980s it became evident that the population was in decline, but the reasons for the decline were unknown. Since that time, the number of bats was further reduced by the hurricanes of 1987, 1990, and 1991: surveys in 1993 recorded only five, and a survey in 1998 found only two. Where have they gone? While the exact reason for their decline remains unknown, scientists speculate that many of the bats perished when hurricane waves surged into their oceanside cave dwellings. After the storms, coral rubble and large logs were packed into the caves, and the walls appeared to have been washed clean.

Numbers of sheath-tailed bats in the western islands of Sāmoa have also declined, but a few small populations have been observed since the hurricanes.

The sheath-tailed bat and the white-rumped swiftlet are frequently confused as being the same animal, and therefore the Sāmoan name **pe'apea'a** is sometimes incorrectly applied to the bat. Confusion probably arises because (1) both animals are very small, swift, agile insect-eaters constantly in motion in search of their prey, and (2) both occupy cave habitat. Numbers of bats may be exaggerated because of confusion with the more abundant white-rumped swiftlet. Nevertheless, it is clear that sheath-tailed bats are declining in number and in danger of extirpation in the archipelago.

MARINE MAMMALS

At least three species of marine mammals listed on the U.S. Endangered Species List—the humpback, sperm, and sei whales—occur in the waters surrounding the archipelago. Other species reported in the archipelago's waters include the blue, fin, and southern right whales. All mammals in U.S. waters are protected by the Marine Mammal Protection Act.

Humpback whales are the most common whale in Sāmoan waters and are typically sighted annually, July through January. September and October are the best months for catching a glimpse of these giants offshore.

Humpback whales living in the Southern Hemisphere number only about 2,500, down from an original estimated population of 100,000 before the days of whaling. Whales living in the Southern Hemisphere are further divided into six groups called "tribes." The small numbers of humpback whales seen in our waters are part of group five. These whales feed in the krill-rich waters off Antarctica during the Austral summer. While many group five humpbacks birth their calves in the waters of Tonga, observations of both adult males and females with calves in the waters of the Sāmoan Archipelago suggest that some humpbacks migrate to our waters to mate and give birth as well.

Spinner dolphins (*Stenella longirostris*) are often seen during boat trips around islands in the archipelago, swimming just off the bow of the boat. They get their name from their acrobatic behavior of jumping out of the water and spinning rapidly on their longitudinal axes (up to seven times) before splashing back down into the ocean. No one knows why they engage in this behavior. Some scientists speculate it is a form of communication; others suggest it is sheer exuberance or play. Spinner dolphins can occur in herds of 1,000, but herds of 200 or fewer are more common. In the Eastern Pacific, spinner dolphins frequently swim above schools of yellowfin tuna. This association is, unfortunately, exploited by the tuna industry. The dolphins, and associated tuna, are herded by high-speed boats, then encircled with huge seine nets. The nets become traps, created by the pursing of the net bottom hanging well below the surface. Unfortunately, the trapped and confused spinners are not always released successfully from the net; tragically, dolphin mortality can sometimes be very high. Dolphin population numbers have been significantly reduced in the last few decades by the tuna purse seine industry. Other species such as whales and manta rays also get trapped in nets.

The sense of sound is greatly evolved in marine mammals. Sound waves travel five times as fast underwater as in air, and travel much greater distances as well. Some species can hear frequencies three times lower, and ten times higher, than humans can. It is thought that marine mammals "hear" sounds primarily using their jawbones instead of their ears. Dolphins and whales also emit sounds (echolocate) to learn about their surroundings. Because sound can penetrate tissues, it is speculated that dolphins can virtually "see" inside each other's bodies and thereby determine the health and well-being of their companions: are they sad? happy? sick? pregnant?

Humpback whale, *Megaptera novaengliae* (16.0 m), IUCN Red List: VU
iʻa manu ("fish animal"), **tafola**

Long-snouted spinner dolphin *Stenella longirostris* (approximately 2.0 m)
mumua, manua

NONNATIVE MAMMALS

Surveys of the islands of American Sāmoa in 1976 reported the Polynesian rat (*Rattus exulans,* **ʻisumu, ʻiole, ʻimoa**) as present on all islands. It was the most abundant of all rat species present, and no doubt is present on the western islands of Sāmoa. It is likely this species has been present since Polynesians first occupied the archipelago, circa 1100 B.C. In American Sāmoa, the house mouse (*Mus musculus,* **ʻisumu, ʻiole**), roof (black) rat (*Rattus rattus,* **ʻisumu**), and Norway rat (*Rattus norvegicus,* **ʻisumu**) have only been observed on islands that have a seaport (Tutuila, ʻAunuʻu, and Ofu). Thus, some species have probably spread to islands with new seaports (e.g., Taʻu, Savaiʻi). The Polynesian rat was even present on Rose Atoll as early as the 1920s. Here, the rats were preying on turtle hatchlings and bird eggs.

Because Rose Atoll is an important seabird rookery and nesting site for sea turtles, scientists began a rat eradication project. Recent surveys indicate that all rats have successfully been removed.

Island birds, in particular, are very vulnerable to predation by introduced mammals such as rats, cats (**pusi**—also the island name for eels), and mongooses (not found in the Sāmoa Islands). Unlike birds living on large continents, island birds often do not have behavioral adaptations that let them coexist with nonnative introduced predators. For example, some species may be flightless, or nearly so, rendering them easy prey. In places like New Zealand, cats have been responsible for the extinction of several island bird species. Rats not only prey on birds, their eggs, and chicks, but may also negatively impact the entire ecosystem and compete with native wildlife for food by eating plants, seeds, fruits, seedlings, and invertebrates. Although there may be other negative factors present, roof rats are most frequently cited as the reason for bird decline or extinction and other island catastrophes.

The mongoose, native to India, was introduced to the Hawaiian Islands to control nonnative rat populations. But this introduction backfired, as many introductions of nonnative species do. The rat is primarily nocturnal, while the mongoose is active during the day. Instead of eating rats, the mongoose has contributed to the decline of many native Hawaiian species, including birds, while the rats continue to flourish.

In the Fijian archipelago, the mongoose (also introduced) has exterminated the buff-banded rail (*Gallirallus philippensis*, **ve‘a**) on several islands.

Feral Pigs

"Feral" is the term applied to otherwise domesticated animals that have reverted to a wild state when they have been released or have escaped from captivity. Recently, large numbers of pigs escaped from pens during the hurricanes of 1987, 1990, and 1991. Feral pigs (*Sus scrofa*, **puaʻa**) are widespread on the various islands, from the coast to the mountaintops, and are multiplying quickly. They can be seen in almost every habitat, including plantations, and lowland and montane rain forest. Feral pigs are also found in almost every watershed in the archipelago. Unfortunately, they root, trample, and destroy native vegetation. They accelerate erosion and pollute our streams with eroded soils, pig feces, and disease. This matter is carried by the streams to the coast, where it is deposited on the reefs. The pigs are thereby damaging the reefs, the fisheries, and the ability of the rain forest to regenerate. In some locations, large areas of forest have been rooted up. This completely destroys the forest floor from revegetation by native trees or plants. In addition, the pigs create muddy wallows where mosquitoes (**namu**) can breed. Some mosquitoes on other islands in the Pacific (e.g., Hawaiʻi) carry a disease known as "avian malaria." This disease is thought to be responsible for the devastation of many native forest bird species. Feral pigs also eat nestlings of ground-nesting birds and the young of other species. In 1998, National Park Service staff from the National Park of American Sāmoa, with the help of staff from Haleakalā National Park (Maui, Hawaiʻi), began a feral pig eradication program on Tutuila. The goal of this ongoing effort is the eradication of feral pigs on park lands.

Feral species in the archipelago also include cats and dogs (**maile**). Unfortunately, escaped pets and livestock are not the only problem. Even well-fed housecats are responsible for the demise of native bird populations in many locations.

BIRDS
(PHYLUM CHORDATA, CLASS AVES)

In number of species, birds are the most diverse group of terrestrial vertebrate wildlife in the Sāmoa Islands. Nevertheless, species diversity is low compared with that observed on larger islands or landmasses. Fifty-four species are listed for American Sāmoa, although only 39 are resident breeders. The rest are migrants or rare visitors, or are observed offshore. An additional 14 species not observed in American Sāmoa regularly occur in the western islands of Sāmoa, for a total of 68 species. In addition, five nonnative species have become "residents."

The birds of the archipelago can be grouped into three broad categories: seabirds, shorebirds, and land birds. The species can be further classified as resident breeders, migrants, rare visitors, and nonnative (i.e., introduced) species. The species described here are resident breeders unless stated otherwise.

SEABIRDS AND SHOREBIRDS

Most of the land birds and shorebirds are well known in the Sāmoa Islands. In contrast, several seabird species in the archipelago are not well known to either Sāmoans or scientists, because their way of life does not bring them to the same areas humans frequent. Seabirds need land for nesting and the ocean for food, typically small fish and squid. Some seabird species nest only on the low coral atolls, while others nest only on high volcanic islands. Still other species are known to nest in both environments. While many seabird species can easily be seen feeding just offshore, some are pelagic and come to land only to breed. Pola Island, located on the north coast of Tutuila, Maga Point on Olosega, and the Aleipata island group, off eastern 'Upolu, are good places to observe (but please do not disturb seabirds in their nesting colonies). Petrels, shearwaters, and storm petrels are the least known, because they are not abundant and their nests are hidden in isolated, mountainous areas such as the top of Mount Lata on the island of Taʻu. These species have more or less been forced to nest in these remote locations because of humans. On islands where there are no introduced nonnative predators or people (e.g., the Rose Atoll of American Sāmoa, now that nonnative rats have been eradicated), seabirds are typically abundant and are found nesting virtually everywhere.

Most seabird species are present year-round (e.g., white tern, brown noddy, white-tailed tropicbird), but occur in greater numbers during the dry, southern hemisphere winter months of April through September. Conversely, migratory shorebirds, most of which nest in the northern tundra, are most abundant during the wet summer months from October to March. The most common shorebird visitor is the Asian golden plover (*Pluvialis fulva*). Unconfirmed resident species and rare visitors are listed in Appendix I.

Island birds are extremely vulnerable to introduced mammalian predators such as rats and cats. Unlike birds living on large continents, island birds often do not have behavioral adaptations that let them coexist with nonnative introduced predators. For example, some species may be flightless, or nearly so, rendering them easy prey.

The Polynesian rat (*Rattus exulans*), accidentally introduced to Rose Atoll, was responsible for the decimation of much of the atoll's nesting seabird populations. The American Sāmoa Department of Marine and Wildlife Resources and the U.S. Fish and Wildlife Service began a rat eradication project in 1990, and current surveys indicate that all rats have now been removed from the islets. Seabirds can now nest without harm to their eggs or themselves. Further, scientists have also recently been successful at eradicating nonnative plant species from the islets. Great care is taken by scientists visiting the atoll not to accidentally reintroduce nonnative plants by inadvertently dropping seeds from personal belongings and effects used in other island locations throughout the Pacific or elsewhere. For example, every item of clothing (as well as all other effects, such as backpacks or tents) worn on or brought to the island must be brand new or previously worn only at Rose Atoll.

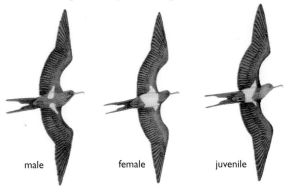

     female   juvenile

Lesser frigatebird, *Fregata ariel* (body length 76 cm/30 in., wing length 183 cm/72 in.)
atafa

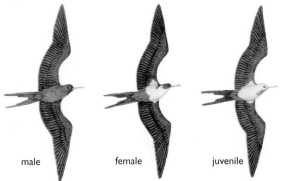

male   female   juvenile

Great frigatebird, *Fregata minor* (body length 94 cm/37 in., wing length 218 cm/86 in.)
atafa

All frigatebirds are kleptoparasites: they steal their food from other birds. While they do catch their own fish, they seem to prefer to harass other birds until their victims drop or even regurgitate their own meals.

Herald petrel, *Pterodroma heraldica* (also classified as species *arminjoniana*)
taʻiʻo (body length 38 cm/15 in., wing length 91 cm/36 in.)

Tahiti petrel, *Pterodroma rostrata* (also classified as genus *Pseudobulweria*) (body length 38 cm/15 in., wing length 84 cm/33 in.)
taʻiʻo

Audubon's shearwater, *Puffinus lherminieri* (body length 30 cm/12 in., wing length 69 cm/27.2 in.)

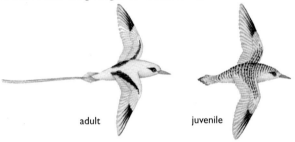

adult   juvenile

White-tailed tropicbird, *Phaethon lepturus* (body length 78 cm/31 in., wing length 91 cm/36 in.), sexes similar
tava'e, tava'e sina

adult   juvenile

Red-tailed tropicbird, *Phaethon rubricauda* (body length 78 cm/31 in., wing length 107 cm/42 in.)
tava'e ula

This tropicbird is uncommon in the western islands of Sāmoa; Rose Atoll is the only known breeding location.

Masked booby (or blue-faced booby), *Sula dactylatra* (body length 86 cm/34 in., wing length 152 cm/60 in.)
fuaʻō
American Sāmoa only; this booby nests only at Rose Atoll.

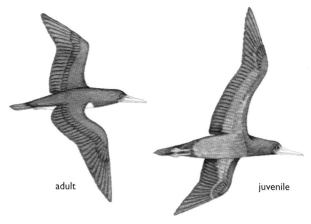

adult

juvenile

Brown booby, *Sula leucogaster* (body length 69 cm/27 in., wing length 142 cm/56 in.)
fuaʻō

adult, white morph   adult, brown morph

Red-footed booby, *Sula sula* (body length 71 cm/28 in., wing length 152 cm/60 in.)
fuaʻō

Pacific reef heron, *Egretta sacra* (body length 58 cm/23 in.)
matuʻu
White morphs and blotched intermediately colored individuals are occasionally seen, especially at Swains Island and Rose Atoll.

Pacific golden plover (or Asian golden plover), *Pluvialis fulva* (body length 27 cm/10.5 in., wing length 44 cm/17.3 in.)
tulī
This plover is a regular migrant visitor.

Ruddy turnstone, *Arenaria interpres* (body length 24 cm/9.5 in.)
tulī, alomalala
The ruddy turnstone is a regular migrant visitor.

Bristle-thighed curlew, *Numenius tahitiensis* (body length 43 cm/17 in.), IUCN Red List: VU
tulī
This curlew is a regular migrant visitor.

Wandering tattler, *Heteroscelus incana* (also classified as genus *Tringa*) (body length 28 cm/11 in.)
tulī, tulī alomalala

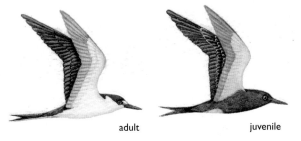

adult   juvenile

Sooty tern, *Sterna fuscata* (body length 43 cm/17 in., wing length 94 cm/37 in.)
gogo uli

Common white (fairy) tern, *Gygis alba* (body length 30 cm/12 in., wing length 86 cm/34 in.)
manu sina

Spectacled (grey-backed) tern, *Sterna lunata* (body length 38 cm/15 in.)
gogosina
Rose Atoll is the only known breeding location in the archipelago.

Blue-gray noddy (or blue noddy), *Procelsterna cerulea* (body length 28 cm/11 in., wing length 61 cm/24 in.)
laia

adult   juvenile

Black noddy, *Anous minutus* (body length 38 cm/15 in., wing length 72 cm/28.3 in.)
gogo

adult   juvenile

Brown (common) noddy, *Anous stolidus* (body length 43
cm/17 in., wing length 86 cm/34 in.)
gogo

Grey duck (or Pacific black duck), *Anas superciliosa* (body
length 56 cm/22 in.); **toloa**

Banded rail (or buff-banded rail), *Gallirallus philippensis*
(also classified as genus *Rallus*) (body length 31 cm/12 in.)
ve'a

White-browed crake, *Porzana cinerea* (also classified as genus *Poliolimnas*) (body length 18 cm/7 in.), western islands of Sāmoa only
vai

Spotless crake (or sooty rail), *Porzana tabuensis tabuensis* (body length 15 cm/6 in.)
This secretive bird, which has no Sāmoan name, is rare in American Sāmoa, where it was last recorded from the island of Ta'u.

Purple swamphen, *Porphyrio porphyrio* (body length 43 cm/17 in.)
manu ali'i, manu sa

LAND BIRDS

The land birds and marsh birds of the Sāmoas occur from the coastal shores up to the tallest mountaintops. Thirty-two land bird species are native to the archipelago; 14 of these occur only in the western islands of Sāmoa. An additional 5 species of land birds are introduced, or "nonnative," species. One well-known nonnative species is a domesticated bird, the chicken, or red jungle fowl (*Gallus gallus*). Small numbers of nonnative, domesticated ducks and geese occur, as do escapees from captivity such as the domestic pigeon, i.e., rock dove (*Columba livia*). One alleged "migrant," the long-tailed cuckoo, or koel (*Eudynamis taitensis*), is seen occasionally throughout the year. All other visitors to date have been either shorebirds or seabirds.

Some species of land birds play an extremely important role in rain forest, and even human, health. Those that are frugivorous (fruit-eating) spread the seeds of the fruiting trees to different areas of the forest in their droppings. Those that are nectarivorous (nectar-drinking) may be essential in pollinating many flowering plants and trees. The insectivorous white-rumped swiftlet (*Collocalia spodiopygius spodiopygius*) consumes hundreds of insects—including insect pests like mosquitoes—each day.

Some species play a role in the Sāmoan culture and are hunted for ceremonial purposes or for consumption during certain holidays. Customs such as these are probably practiced less frequently due to a dwindling population. Pacific pigeons (**lupe**) and white-throated pigeons (**fiani**) are occasionally hunted for food, particularly for Easter and White Sunday holidays. Substantial declines in their numbers have occurred after hurricanes. While some mortality may occur directly as a result of the hurricane itself, scientific evidence suggests that much mortality is due to increased hunting vulnerability after hurricanes, when the native forest is stripped of leaves and fruit, and birds must forage widely and in open habitats in search of food. In response to the recent hurricanes, both nations imposed a hunting ban to allow the species time to recover. Unfortunately, the hunting still occurs in American Sāmoa, and the ban is not being effectively enforced in the western islands of Sāmoa either.

The stories of the village chiefs tell of past days when birds were plentiful.

Endemic Birds

Ten species of land birds are endemic to the archipelago, including two genera, *Pareudiastes* and *Didunculus*. Sadly, the Sāmoan woodhen, or wood rail (*Pareudiastes [Gallinula] pacificus*, **puna'e**), endemic to the island of Savai'i in the western islands of Sāmoa, was last recorded with certainty in 1873 and was, until recently, believed to be extinct. Nevertheless, recent possible sightings in the area of Mount Elietoga in 1987 give hope that small numbers may survive in remote locations (IUCN *Red List of Threatened Animals*). The tooth-billed pigeon (*Didunculus strigirostris*, **manumea**) is endemic to the western islands of Sāmoa. First called the "little dodo" by scientists, it caused great confusion because of the resemblance of its bill to that of the now-extinct dodo (which actually was a giant pigeon). Although it has been declared the national bird, the government has also declared it in danger of extinction. Of the seven additional species endemic to the archipelago, one, the Sāmoan starling (*Aplonis atrifusca*, **fuia**) is found throughout the islands, and six are found only in the western islands of Sāmoa. Six subspecies are endemic to American Sāmoa, five to the western islands of Sāmoa. A subspecies is a subdivision of a taxonomic species, and is usually limited to a specific geographic area. The collared kingfisher (*Halcyon chloris manuae*) and the Fiji shrikebill (*Clytorhynchus vitiensis powelli*) of the Manu'a Islands are two such subspecies.

Nonnative Birds

Nonnative bird species can totally disrupt native bird populations. Since all species can be disease carriers, introduced birds may be reservoirs for diseases to which native species have no tolerance, such as avian malaria. Native species can be exterminated by such diseases. Further, some nonnative species are very aggressive and displace native birds from their territories or outcompete them for food resources or nest sites.

One common nonnative bird, often domesticated but also occurring in the forest, is the red jungle fowl, *Gallus gallus*, **mao'aivao, tama'i moa, toa** (not illustrated). Native to Asia, it is frequently known as a "chicken."

Red-vented bulbuls are highly aggressive birds from Southeast Asia. Introduced to the Sāmoa Islands, they are now found locally near human habitation, in plantations, etc. They displace other species from their territories and outcompete them for food. They also raid gardens, taking bites out of everything. At one time, wattled honeyeaters were common around household dwellings and plantations. Because of the introduction of the bulbul, in some areas it is now rare to have a wattled honeyeater in the yard.

The jungle myna and common myna, introduced to American Sāmoa, are reproducing and spreading rapidly. Mynas, like bulbuls, are aggressive and displace native birds. Mynas may compete with the endemic Sāmoan starling (**fuia**) for nesting cavities. Fields that were once filled with overwintering Asian golden plovers (**tulī**), putting on fat reserves for the long migration north to their breeding grounds in the Northern Hemisphere, now must be shared with noisy, annoying mynas. How adaptable are the starlings and plovers to these new species? Do they have somewhere else to go?

Many-colored fruit dove (or rainbow dove), *Ptilinopus perousii perousii* (body length 22 cm/8.5 in.), crimson under tail; female similar in coloration to purple-capped fruit dove

manumā, manulua

This subspecies is endemic to the archipelago and locally rare throughout. The **manumā** is primarily seen feeding on the fruits of banyan (**aoa**) trees and is less often observed now that these trees have been cut down to make way for plantations and housing. Furthermore, surveys show that its numbers decreased after the hurricanes of 1987, 1990, and 1991, when the remaining banyan trees were stripped of their leaves and fruits and hunting pressure increased.

Purple-capped fruit dove (or crimson-crowned fruit dove), *Ptilinopus porphyraceus* (body length 23 cm/9 in.), yellow under tail, compare with many-colored fruit dove; male and female similar

manutagi (adult), **manufili** (young)

This bird is frequently heard in plantations and near dwellings, but less often seen. Numbers of this species also declined significantly after the hurricanes of 1987, 1990, and 1991.

Pacific pigeon (Pacific imperial pigeon), *Ducula pacifica microcera* (body length 41 cm/16 in.)
lupe

Friendly ground dove (or shy ground dove), *Gallicolumba stairi* (body length 25 cm/10 in.), IUCN Red List: VU (imminently)
tu'aimeo, tiotala

Long-tailed cuckoo (or long-tailed koel), *Eudynamis taitensis* (body length 41 cm/16 in.)
Recent data suggest that this species, once thought to be an austral winter visitor, may be seen infrequently throughout the year.

Blue-crowned lorikeet, *Vini australis* (body length 18 cm/7 in.)
sega, **sega'ulu**, **sega Sāmoa** (American Sāmoa; most often observed in Manu'a); **sega vao** (western islands of Sāmoa)

White-collared kingfisher (or collared kingfisher), *Halcyon chloris* (also classified as genus *Todirhamphus*) (body length 22 cm/8.5 in.)
ti'otala
The subspecies *H.c. pealei* is endemic to the island of Tutuila; *H.c. manuae* is endemic to the Manu'a Islands; American Sāmoa only.

White-rumped swiftlet, *Collocalia spodiopygius spodiopygius* (also classified by some scientists as genus *Aerodramus*) (body length 10 cm/4 in.)
pe'ape'a

This subspecies is endemic to the archipelago. It nests in caves in total darkness, where it navigates using an echolocation system similar to that used by many bats. Its nests are constructed using mosses, lichens, and fine leaves glued together with the swiftlet's own saliva.

Wattled honeyeater, *Foulehaio carunculata* (body length 18 cm/7 in.)
iao

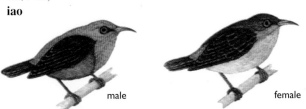

male   female

Cardinal honeyeater (or cardinal myzomela), *Myzomela cardinalis* (body length 13 cm/5 in.)
segasegamau'u
This species is absent from the Manu'a Islands

Tutuila, western islands of Sāmoa   Manu'a Islands

Polynesian starling, *Aplonis tabuensis,* (body length 18 cm/7 in.)

miti, **miti vao** (American Sāmoa); **mitiula** (western islands of Sāmoa)

According to Amerson et al. (1982), the *subspecies A. t. tutu-ilae* is endemic to the islands of Tutuila and 'Aunuu, *A. t. manuae* to Manu'a, and *A. t. brevirostris* to the western islands of Sāmoa.

Sāmoan starling, *Aplonis atrifusca* (body length 25 cm/10 in.)
fuia
This starling is endemic to the Sāmoa Islands.

Fiji shrikebill, *Clytorhynchus vitiensis powelli* (body length 18 cm/7 in.)
segaolevau
This subspecies is endemic to the Manu'a Islands; American Sāmoa only

Barn owl, *Tyto alba* (body length 41 cm/16 in.)
lulu
This species is found worldwide.

Tooth-billed pigeon, *Didunculus strigirostris* (body length
31 cm/12 in.), IUCN Red List: VU
manumea
Western islands of Sāmoa only

White-throated pigeon, *Columba vitiensis castaneicieps*
(body length 41 cm/16 in.)
fiaui
Western islands of Sāmoa only

Flat-billed kingfisher, *Halcyon recurvirostris* (also classified as genus *Todirhamphus*) (body length 18 cm/7 in.)
ti'otala
Western islands of Sāmoa only

Sāmoan broadbill (or Sāmoan flycatcher), *Myiagra albiventris* (body length 15 cm/6 in.), IUCN Red List: VU
tolaiʻula, **tolai fatu**
Western islands of Sāmoa only

Island of ʻUpolu

Island of Savaiʻi

Sāmoan fantail, *Rhipidura nebulosa* (body length 14 cm/5.5 in.)
sau
The subspecies *R. h. altera* is endemic to the island of Savaiʻi, and *R. h. nebulosa* to ʻUpolu.
Western islands of Sāmoa only

Polynesian triller, *Lalage maculosa* (body length 15 cm/6 in.)
Western islands of Sāmoa only

Sāmoan triller, *Lalage sharpei* (body length 13 cm/5 in.),
IUCN Red List: VU
miti tae
The subspecies *L. s. tenebrosa* is found above 600 m on the
island of Savai'i, and *L. s. sharpei* above 200 m on 'Upolu
in the Western islands of Sāmoa only

Island thrush, *Turdus poliocephalus samoensis* (body length
23 cm/9 in.)
tutumalili
Western islands of Sāmoa only

Sāmoan white-eye, *Zosterops samoensis* (body length 10
cm/4 in.), IUCN Red List: VU
mata papae

This bird is endemic to the island of Savai'i; it is almost always found at elevations above 900 m.; western islands of Sāmoa only

male female

Scarlet robin, *Petroica multicolor* (body length 10 cm/4 in.)
tolaiula
Western islands of Sāmoa only

Sāmoan whistler, *Pachycephala flavifrons* (body length 15 cm/6 in.)
vasavasa
This bird is endemic to the western islands of Sāmoa and found only there.

Red-headed parrot finch, *Erythrura cyaneovirens cyaneovirens* (body length 10 cm/4 in.)
manu ai pau laau
Western islands of Sāmoa only

Mao, *Gymnomyza samoensis* (body length 31 cm/12 in.),
IUCN Red List: VU
maomao
This bird was last seen in American Sāmoa in 1977; currently known from western islands of Sāmoa only.

Nonnative Land Birds

Jungle myna, *Acridotheres fuscus* (body length 23 cm/9 in.). Native to Asia.

Common myna, *Acridotheres tristis* (body length 25 cm/10 in). Native to Asia.

Red-vented bulbul, *Pycnonotus cafer* (body length 22 cm/8.5 in.). Native to Asia.
manu pālagi, manu papālagi
Western islands of Sāmoa only

Rock dove or domestic pigeon, *Columba livia* (body length 32 cm/12.5 in.). Common in Apia on the western island of 'Upolu.

TABLE I.
RARE OR UNCONFIRMED RESIDENT BIRD SPECIES AND VISITORS
(* DENOTES NONBREEDING):

Whimbrel,* *Numenius phaeopus* (body length 46 cm/17 in.)

Bar-tailed godwit,* *Limosa lapponica* (body length 41 cm/15.5 in.)

Grey-tailed tattler, *Tringa brevipes* (body length 27 cm/10.5 in.)

Sanderling,* *Calidris alba* (body length 21 cm/8 in.)

White-faced heron,* *Ardea novaehollandiae* (body length 66 cm/26 in.)

Cattle egret,* *Bubulcus ibis* (body length 51 cm/20 in.)

Great crested tern,* *Sterna bergii,* **tala** (body length 48 cm/19 in., wing length 109 cm/43 in.)

Bridled tern,* *Sterna anaethetus* (body length 38 cm/15 in., wing length 76 cm/30 in.)

Black-naped tern,* *Sterna sumatrana* (body length 32 cm/12.5 in., wing length 61 cm/24 in.)

Phoenix petrel, *Pterodroma alba* (body length 35 cm/14 in., wing length 83 cm/32.5 in.)

Collared petrel, *Pterodroma (leucoptera) brevipes* (body length 30 cm/12 in., wing length 71 cm/28 in.)

White-necked petrel,* *Pterodroma externa*, also classified as *Pterodroma cervicalis* (body length 43 cm/17 in., wing length 91 cm/36 in.)

White-throated storm petrel, *Nesofregetta fuliginosa* (body length 26 cm/10 in.)

White-faced storm petrel, *Pelagodroma marina* (body length 20 cm/8 in., wing length 42 cm/16.5 in.). A wing of this species washed ashore in American Sāmoa in 1992.

Newell's shearwater (also known as Townsend's shearwater), *Puffinus auricularis newelii*. An injured bird was found in American Sāmoa in 1993; it is known to breed only in Hawai'i .

Sooty shearwater, *Puffinus griseus* (body length 46 cm/18 in., wing length 104 cm/41 in.). Two injured birds have been found in American Sāmoa, in 1992 and 1994.

Short-tailed shearwater (also known as slender-billed shearwater),* *Puffinus tenuirostris* (body length 43 cm/17 in., wing length 100 cm/39.5 in.).

Wedge-tailed shearwater, *Puffinis pacificus* (body length 38 cm/15 in., wing length 81 cm/32 in.). This species was heard calling (but not seen) in 1994 at Rose Atoll; it may start breeding there again now that the rats have been eradicated.

Christmas shearwater, *Puffinis nativitatus*. This species was observed at Rose Atoll in 1994; it may start breeding there again now that the rats have been eradicated.

Laughing gull, *Larus atricilla* (body length 43 cm/17 in., wing length 107 cm/42 in.)

REPTILES AND AMPHIBIANS

(GECKOS, SKINKS, SNAKES, AND SEA TURTLES)

Reptiles and amphibians are collectively known as "herpetofauna." Twenty-three species of reptiles (phylum Chordata, class Reptilia) and a single species of amphibian (phylum Chordata, class Amphibia) are recorded for the archipelago (Table II). Sixteen species of reptiles and the single amphibian are terrestrial; five turtles and two sea snakes are strictly marine (except for times of egg laying). Most species of reptiles hatch from eggs and are miniature replicas of the parents. The tiny (1 cm) eggs of several gecko species are commonly found tucked in crevices around human dwellings.

The Sāmoan skink (*Emoia samoensis*) is endemic to the islands of Sāmoa. One species of skink, Murphy's skink (*Emoia murphyi*), is recorded only for the western islands of Sāmoa, while Lawe's skink (*Emoia lawesii*) is recorded only for American Sāmoa. A few species of geckos now present in the archipelago are thought to have been introduced rather recently. The only amphibian, the marine toad, was introduced to American Sāmoa in 1953 (see below). Amphibians as a whole are not well represented on oceanic islands because their soft, sensitive, and highly permeable skin does not lend itself to dispersal across salt water.

REPTILES
Geckos

There are approximately 850 species of geckos (**moʻo**) worldwide; five species are recorded from the Sāmoan Archipelago. Most geckos are nocturnal and have loud voices that you can hear as they chirp back and forth. Their eyelids are immovable, and so their eyes are always open. The undersides of their enlarged toe pads contain thousands of Velcro®-like bristles, or setae. Each bristle has its own microscopic suction cup at the end. These suction cups allow the gecko to climb vertical surfaces and walk across almost anything, including windows, walls, and ceilings. Most species lay clutches of two leathery eggs that harden upon drying.

As a survival mechanism, most geckos are capable of "discarding" their tails to escape being eaten. If a predator grabs the tail, it simply falls off, thereby permitting the now tailless gecko to escape. Also, the shade of their skin color changes to match their background. The common house gecko, brown-speckled in the daytime, is often a creamy tan or pink when feeding near lights at night. Geckos are extremely beneficial neighbors and houseguests. They consume tremendous quantities of pest insects, such as cockroaches, termites, and mosquitoes.

House gecko, *Hemidactylus frenatus*
moʻo
The house gecko is believed to be a relatively recent accidental introduction to American Sāmoa. This nocturnal gecko is found only on or near buildings and thrives around human dwellings, especially near outside light fixtures, where it feeds on the abundance of insects attracted by the lights. The intensity of its typical coloration varies from a dark creamy gray during the day to a pale shade of peach or gray at night. In contrast to mourning geckos (described below), which are unisexual, house geckos are both male and female.

Mourning gecko, *Lepidodactylus lugubris*
mo'o
The mourning gecko is the smallest, yet least timid gecko
species in the archipelago and is found in a variety of habitats, including secondary forest and plantations. Like the
house gecko, it most commonly occurs in close association
with humans and their structures. It is nocturnal and hides in
cracks and crevices during the day, emerging at night to feed.
It appears that mourning geckos are displaced from crevices
occupied by house geckos. Interestingly, mourning geckos
are parthenogenic: all individuals in the population are sin-
gle-sexed (unisexual). In this case, all are female and produce fertile eggs. In addition, this species is a communal
nester, and several females may lay their eggs in the same
location. Its coloration is a speckled brown background with
W-shaped crossbands on its back.

Stump-toed gecko, *Gehyra mutilata*
mo'o

The stump-toed gecko is an expert at camouflage; its color changes to blend into its environment. This nocturnal and arboreal gecko has particularly wide toe pads, and the base of its tail is often constricted. The stump-toed gecko inhabits buildings of rather recent origin, capitalizing on insects attracted by lights, and is present in adjacent vegetation or forest. In contrast to mourning geckos, described above, which are unisexual, stump-toed geckos are both male and female.

Nonnative Species

Few nonnative animals end up being completely harmless to their new environments, especially in island ecosystems, and many devastate the native species. For example, mourning geckos may become increasingly rare in American Sāmoa because of the introduction of two nonnative species: the house gecko and the red-vented bulbul. House geckos are pugnacious hunters and can outcompete mourning geckos for insects. Further, in Hawai'i, the red-vented bulbul has been feeding heavily upon the mourning gecko and decimating its numbers.

The mongoose, an introduced species, has decimated the terrestrial skink population on the Fijian island of Viti Levu.

Skinks

There are over 1,000 species of skinks (**pili**) worldwide. Nine species are known from the Sāmoan Archipelago, although their taxonomy is confusing and still under debate. Most species are diurnal and feed on a variety of insects, including centipedes. Some skinks (e.g., the brown-tailed copper-striped skink, *Emoia cyanura*) are extremely abundant, reaching a density of perhaps 1,000 individuals per hectare.

Like geckos, skinks have tails that detach, enabling them to evade capture by a predator. While the tail does regrow in several months, the replacement is usually somewhat stunted and less colorful than the original. The illustrated species of skinks are diurnal and frequently encountered either around human dwellings or in forested habitats.

Sāmoan skink, *Emoia samoense*
pili lape
This species is endemic to the archipelago. It is unusual in that it is a semi-arboreal species. It spends a lot of time on trees, but also forages on shrubs and on the ground at the base of trees. It appears to be restricted to forested habitat, so encountering one near houses is unlikely.

Blue-tailed copper-striped skink, *Emoia impar*
pili
This is a diurnal skink that frequents forested habitats as well as human settlements, plantations, and forest edges. While it is most active when the sun is shining, it does move about on overcast days as well. It will climb all types of vegetation looking for food, but rarely ventures to a height exceeding 1

m (approximately 3 ft.). The blue-tailed copper-striped skink and brown-tailed copper-striped skink (*Emoia cyanura*) are sibling species. This means that they are nearly identical in outward appearance but do not breed with each other. The tail of the blue-tailed copper-striped skink is often bright blue (rarely brownish), and the belly and underside of the thighs are a dusky color. Conversely, the tail of the brown-tailed copper-striped skink is usually brownish, although it can occasionally be a pale blue-green (but never bright blue). The belly and undersides of its thighs are white. See Zug (1991) for additional species-specific characteristics.

Pacific black skink, *Emoia nigra*
pili uli
While this skink can be observed in village habitats, it is most abundant in forested habitats. This diurnal skink is rarely seen on overcast days, but is highly active on sunny ones.

Snake-eyed skink, *Cryptoblepharus poecilopleurus* (formerly known as *C. boutonii*)
pili

This species is restricted to rocky coastal habitat, living on rocks, beach rubble, and the vegetation located just above the high-tide line. This species is not recorded from the islands of Tutuila, 'Aunuu, and Ofu. Snake-eyed skinks have immovable eyelids, which make their eyes appear larger. You might catch a glimpse of them scurrying into the labyrinth of holes in volcanic boulders on the shoreline if you get too close.

Sea Turtles

Five species of sea turtles (**laumei**) are recorded for the archipelago. These species include the federally endangered hawksbill (*Eretmochelys imbricata*) and leatherback (*Dermochelys coriacea*) sea turtles and the federally threatened green (*Chelonia mydas*), loggerhead (*Caretta caretta*), and olive ridley (*Lepidochelys olivacea*) sea turtles. In addition to being on the U.S. Endangered Species List, they are also protected by nations participating in CITES (see below). The green and hawksbill turtles are the most common turtles observed in the archipelago and are recorded in small numbers around all the Sāmoa Islands. The hawksbill sea turtle was placed on the U.S. Endangered Species List as "endangered" in 1970; the green sea turtle was placed on the list as "threatened" in 1978. The other species listed above are considered rare visitors.

Not much is known about sea turtle life history before the turtles reach maturity. In fact, no one knows exactly how long sea turtles live or at what age they begin to reproduce. It is thought that females do not become sexually mature until somewhere between 20 and 50 years of age, depending on the species.

Females usually nest only every second to fourth year. When the moon and tides are just right, the females emerge from the ocean under the cover of darkness. Using both hind flippers to scoop out a deep nest, hopefully above the high-tide line, the female deposits a clutch of 100 eggs or more. After covering the eggs with sand, she returns to the sea, and the nest is on its own. The female will nest an average of three or more times, once every two or more weeks. After that, she will not nest again for several more years.

In approximately two months, if the nest has survived digging up by predators and people, the eggs begin to hatch. Under the cover of darkness, the little turtles break through their sand chamber to the surface and make their way to the sea. It is likely that only a single turtle, sometimes none, out of every 100 that hatch, will successfully avoid predation by sharks, seabirds, fishing nets, and other mortality factors and live to reach maturity.

Upon hatching, the little turtles make a beeline toward the twinkling ambient light cast by the ocean. Unfortunately, streetlights and other illumination can confuse the newborn turtles, causing them to go the wrong way. Artificial light often attracts them dangerously close to roads, cars, and dogs, not to mention away from the ocean. If you find a nest of hatched baby turtles, please return them immediately to the sea—it is their only chance of survival. There is no truth to the tale that their mothers are waiting offshore to eat them.

Outside of the United States and its territories, such as American Sāmoa, various laws protect endangered species such as sea turtles. CITES legally prohibits member nations from engaging in international commerce in species on the CITES species list, but it does not prevent the harassment or exploitation of sea turtles and their nests. Unfortunately, the independent nation of Sāmoa (western islands of Sāmoa) is not a member of CITES. Further, while it is illegal to disturb a sea turtle nest, or sell or destroy sea turtle eggs in the western islands of Sāmoa, it is, sadly, not considered a crime to disturb or kill the turtle itself.

🐟 🐟 🐟 🐟 🐟 🐟 🐟 🐟 🐟 🐟 🐟 🐟 🐟 🐟 🐟 🐟

Seven species of sea turtles are found worldwide, all endangered. Much sea turtle mortality is associated with human activities. As many as 60,000 sea turtles are killed annually because of shrimp fishing and entanglement in other fishing nets and debris. In the Sāmoan Archipelago, erosion of nesting beaches has resulted in loss of suitable nesting habitat. Other historical nesting sites have been ruined with sea walls and rock revetments. Artificial lighting, such as street lights, deters females from nesting and disorients turtle hatchlings as they attempt to make their way toward what they think is the glow of the sea. Locally, sea turtle nests are frequently dug up, both by dogs and people, who eat the eggs. Every time a sea turtle nest is raided, an endangered species is closer to extinction. Some turtles are stuffed as trinkets for sale at markets. Moreover, adult turtles are killed so a few bracelets and earrings of turtle shell can be made. It is more precious to see a live turtle than a dead one.

🐟 🐟 🐟 🐟 🐟 🐟 🐟 🐟 🐟 🐟 🐟 🐟 🐟 🐟 🐟 🐟

Population sizes and nesting locations for green and hawksbill sea turtles on local island beaches are generally poorly known. Turtle nesting activity in the archipelago is thought to occur year-round, but the nesting cycles of each species are not clearly defined. Only very small numbers of the two species have nested on Swains Island. Green turtles do consistently nest at Rose Atoll National Wildlife Refuge, however, typically during August and September, and perhaps less frequently during the rest of the year.

For several years, scientists from the United States National Marine Fisheries Service and American Sāmoa Department of Marine and Wildlife Resources have monitored this nesting population. Several green sea turtles nesting at Rose Atoll have been tagged and fitted with satellite transmitters to see where they go after nesting. As it turns out, sea turtles nesting at Rose Atoll range thousands of miles across the Pacific. The results of the first year's tagging program showed that three turtles went west to the islands of Fiji. The following year, the single tagged turtle went the opposite direction, east to the islands of French Polynesia. Subsequent tagged turtles have again made their way west to Fiji. One of these turtles has been nesting at Rose for over ten years.

In Hawai'i and other sites around the world, a large proportion of green sea turtles are suffering from a new viral disease, usually fatal, that covers the turtles with tumors. Scientists are unsure of the cause, but suspect nonpoint source pollution. Turtles with tumors have not yet been reported in Sāmoan waters.

Why do you think sea turtles come to Rose Atoll to nest each year? Where would they go if Rose Atoll's nesting habitat was ruined? (See Rose Atoll National Wildlife Refuge section in Protected Reserves chapter for a description of the damage done when a Taiwanese ship crashed into the reef at Rose, spilling oil and other lubricants over a large area.)

Do not touch, disturb, or harass sea turtles or their nests. All species of sea turtles are endangered; they are protected by U.S. federal law in American Sāmoa. Greater protection for sea turtles in the western islands of Sāmoa is warranted.

Pacific hawksbill sea turtle, *Eretmochelys imbricata*, IUCN Red List: CR; U.S. Endangered Species List: Endangered
laumei una, iʻasa
This hawksbill subspecies occurs on coral reefs throughout the Pacific. Hawksbill turtles may be hard to find because they are not only timid, but also highly endangered. Hawksbill turtles forage near rock and reef habitats, favoring sponges. They may infrequently consume crustaceans and other animal matter. Note the distinctive bill-like snout.

Pacific green sea turtle, *Chelonia mydas*, IUCN Red List:
EN; U.S. Endangered Species List: Threatened
laumei, lanumeamata
The green sea turtle grows much larger than the hawksbill.
The shell can reach 1.2 m (4 ft.) in length, while the turtle
itself can weigh more than 145 kg (320 lbs.). Unlike their
carnivorous cousins, green sea turtles are primarily herbivorous, feeding on a variety of sea grasses and algae (seaweed).
This turtle's name comes from the green color of its fat, not
the shell, which is usually streaked with dark brown or dark
olive. This species is sometimes commonly referred to as the
black turtle.

Terrestrial and Sea Snakes
    The only terrestrial snake native to the archipelago is
the Pacific boa (*Candoia bibroni*, **gata**). It has been found in
plantation or forest habitat close to the coast or along the
beach strand itself. Although such behavior is rare, a boa was
recently (1997) observed sunning itself on a beach on the
island of Ta'u. This island is the only known location for this

snake in American Sāmoa. While still uncommon, boas may be more abundant on the western islands of Sāmoa.

Approximately 50 species of sea snakes are found worldwide, almost all in the family Hydrophidae. The majority of sea snakes are encountered close to the continental coasts of Asia and Australia, including the Indonesian archipelago.

Two species of sea snakes are recorded for the archipelago, the banded sea krait (*Laticauda* sp., **onea**) and the yellow-bellied sea snake (*Pelamus platurus*). Sea snakes are highly venomous, but rarely aggressive; an encounter (which is unlikely) is no cause for panic. Sea snakes are basically inoffensive and mind their own business. Nevertheless, they should not be handled. Fatalities have been recorded in Southeast Asia, mainly to fishermen removing snakes tangled in fishing nets.

Sea snakes, it should be noted, are frequently confused with snake eels. Unlike eels, sea snakes have scales and paddle-shaped tails. Eels usually have long fins running along their dorsal and ventral surfaces; snakes have no fins. The harlequin snake eel (*Myrichthys columbrinus*) has a pattern almost identical to that of the banded sea krait (see Coral Reef and Nearshore Fishes chapter).

Pacific boa, *Candoia bibroni* (maximum known length is just over 1 m/approx. 45 in.)
gata

Island blind snake, *Ramphotyphlops braminus*

The island blind snake is a harmless, tiny (approximately 15 cm), wormlike, fossorial (living under soil or leaf litter) snake that feeds on small insects and other invertebrates. It has vestigial eyes that appear only as two tiny spots. The blind snake is not native to the archipelago and probably arrived accidentally in potting soil. Color is variable. Individuals observed locally (so far, it has been recorded only for American Sāmoa) have been rusty brown, gray-blue, or coral, with lighter ventral areas. The island blind snake is the only known snake that is parthenogenic (unisexual). All individuals are female, and their eggs develop without fertilization by a male.

Banded sea krait, *Laticauda* sp.
gata sami
Although this snake can easily be confused with several similar-looking snake eels, an individual washed up on Utulei Beach on Tutuila in the late 1980s, thus confirming its presence in the archipelago.

Yellow-bellied sea snake, *Pelamus platurus*

Warning: Potential Environmental Crisis

The brown tree snake (*Boiga irregularis*), native to the islands of New Guinea, was accidentally introduced to the island of Guam sometime after World War II, probably hiding in a shipment on a military cargo plane. Because of a lack of natural predators on Guam, this tree snake has now reached population levels of over 50 snakes/acre in some areas of the island. This represents the highest density of snakes found anywhere in the world. Its presence has become an environmental and economic nightmare. Guam has become an island on which the dawn chorus of birds (**le tausagi**) no longer exists. By raiding bird nests and eating the flightless nestlings, this snake wiped out almost every bird species on the island. When mother bats with dependent young go out to feed, they leave their offspring hanging on branches. The snakes find the baby bats easy prey. Not only has this snake decimated almost all native bird and bat life on the island, but it has negatively affected tourism. Hundreds of power outages on the island occur yearly as the snakes slither over the wires, inducing electrical shorts. In addition, the snakes have been found everywhere, including inside toilets, baby cribs, automobile engines, and the wheel wells of planes.

Because airplanes regularly fly between Guam and Hawai'i, in 1990 the State of Hawai'i formed a Brown Tree Snake Control Group for the express purpose of preventing this snake from colonizing the Hawaiian Islands and destroying the islands' biological diversity and associated tourism economy. Seven snakes have been found and eradicated on or near the Honolulu Airport on the island of Oʻahu. Nevertheless, one snake that was seen at Hickam Air Force Base escaped capture. It would take only one pregnant snake to ultimately create a big problem; nevertheless, there is no known population of brown tree snakes in the Hawaiian Islands at this time.

Military and other flights from around the South Pacific stop regularly in the Sāmoan Archipelago, potentially bearing brown tree snakes. It is important to be aware of this potential nightmare and work accordingly to protect native fauna.

AMPHIBIANS
Nonnative Herpetofauna Species

One nonnative species, the marine toad, was introduced to the island of Tutuila. Fortunately, this toad has not spread to the neighboring islands of Manuʻa or the western islands of Sāmoa.

Marine toad, *Bufo marinus*
lage, rane
The marine toad, native to Central America and northern

South America, was intentionally introduced to Tutuila from Hawai'i in 1953 to control centipedes, mosquitoes, and insects that attack the taro plant. In Australia, it is a well-known nonnative pest called the "cane toad."

Initially released into artificial ponds on the island of Tutuila in the village of Taputimu, the toads have since spread, and occur from the seashore to the top of Mt. Alava. Biologists estimate that there are over two million toads on Tutuila. It is not clear what impact these toads are having on components of the native Sāmoan environment at this time. While they no doubt eat many insects, these same insects were previously destined for the stomachs of native bird species and the very rare insect-eating sheath-tailed bat, **tagiti**. It is also hard to imagine a toad eating a centipede, considering the centipede's fierce nature, savage demeanor, seemingly indestructible exterior, and tenacity to live.

One worry regarding this toad is its large, toxic skin glands. These glands are located just behind the eyes, on the toad's "shoulders." When distressed, the toad secretes a poisonous white fluid from these glands. The poison, called "bufogen," is toxic if consumed, or if rubbed into the eyes, mouth, etc. Dogs have died from mouthing or attempting to eat these toads.

TABLE II.
HERPETOFAUNA (REPTILES AND AMPHIBIANS) OF THE SĀMOAN ARCHIPELAGO

Reptiles

Geckos:
Stump-toed gecko, *Gehyra mutilata*
Oceanic (Polynesian) gecko,** *Gehyra oceanica*
House gecko, *Hemidactylus frenatus*
Mourning gecko,** *Lepidodactylus lugubris*
Pacific slender-toed gecko, *Nactus pelagicus*‡
Skinks:
Snake-eyed skink, *Cryptoblepharus poecilopleurus*
Micronesian skink, *Emoia adspersa* (reported for the western islands of Sāmoa and Swains Island)
Lawe's skink, *Emoia lawesii* (reported for American Sāmoa only)
Murphy's skink, *Emoia murphyi*
Pacific turtle skink, *Emoia nigra*
Sāmoan skink, *Emoia samoensis*
Brown-tailed copper-striped skink, *Emoia cyanura*†
Blue-tailed copper-striped skink, *Emoia impar*†
Moth skink, *Lipinia noctua*

Snakes:
Pacific boa, *Candoia bibroni*
Banded sea krait, *Laticauda* sp.
Yellow-bellied sea snake, *Pelamus platurus*
Island blind snake, *Ramphotyphlops braminus* (reported for American Sāmoa only)

Sea turtles:
Green sea turtle, *Chelonia mydas*

Hawksbill sea turtle, *Eretmochelys imbricata*
Olive ridley turtle, *Lepidochelys olivacea*
Leatherback, *Dermochelys coriacea*
Loggerhead, *Caretta caretta*

Amphibians
Marine toad, *Bufo marinus*

** Also recorded on Rose Atoll.
‡ *Crytodactylus pelagicus* in Amerson, et al.
† *Emoia impar* and *E. cyanura* are very difficult to tell apart in the field.
Amerson et al. (1982) and Schwaner (1979) do not report *E. impar* as
occurring on American Sāmoa, but both species probably do co-occur.
For species-specific morphologic and taxonomic characteristics see text,
and Zug (1991).

TERRESTRIAL INVERTEBRATES

LAND SNAILS

Land snails (**sisi vao**), air-breathing gastropods called "pulmonates," are the terrestrial cousins of the sea snails and nudibranchs. Both belong to the phylum Mollusca. A total of 94 land snails and slugs are found in the archipelago, in addition to 46 freshwater and brackish-water species (Cowie 1998). Seventy-six species of land snails are native to Sāmoa, and 59 of these are endemic. Further, 34 of the 59 endemic species are endemic to single islands (e.g., *Ostodes strigatus* is found only on Tutuila). Several other species of introduced snails are commonly observed near human habitation (e.g., *Subulina octona*, *Bradybaena similaris*).

Land snails are primarily nocturnal. During the day, most snails remain sealed against the underside of leaves, often not far from the ground. Some are sealed against tree trunks or branches; others are found under dead leaves on the ground. If the night is very dry, the snails may remain sealed to their surface. Conversely, if the day is humid or rainy, the snails will move about during daylight. Some snails never wander too far and may even spend their entire lives on a single plant. Snails and slugs have special glands in their bodies that secrete a type of mucus; the snail slides along this slimy trail by contracting its large, muscular "foot." Pulmonates are hermaphroditic, producing both sperm and eggs. Some species are also self-fertilizing, a useful trait for rare species that may have difficuilty locating mates. Those species with separate males and females are mostly marine snails.

Several native land snails of Sāmoa belong to the family Partulidae, in the genera *Eua* and *Sāmoana*. Interestingly, this family of arboreal (tree-based) snails exists only on high islands in the Pacific. Unfortunately, partulids are now highly endangered and therefore rarely seen, although they were at one time very common and widespread. Many factors have led to their rarity. Early Polynesians cleared lowland habitat to make room for dwellings and agricultural plantations. Today, both lowland and mountain forests continue to be cut down. European settlers arrived and brought with them nonnative animals such as rats, which ate the snails and further decreased their populations. Even shell collectors may have reduced snail numbers. One collection currently dangles from the lobby of the Rainmaker Hotel on Tutuila, where the hanging chandeliers are decorated with an estimated 10,000 partulid snail shells, mostly from the precariously rare *Eua zebrina*.

Shell color in live animals can be quite different from that of empty shells. The land snails illustrated below, which were painted from museum specimens, would probably appear darker if observed alive in the field. Sometimes, shells without a live animal lose all coloration and appear almost white.

One helpful identification feature of land snail shells is the position of the shell's aperture (opening). Shells are either "left-handed" or "right-handed." To determine whether a shell is left- or right-handed, hold an empty shell with its pointed end up, aperture facing toward you. The aperture will face either left (left-handed) or right (right-handed). The illustrated shells are all right-handed unless noted otherwise.

Trouble in Paradise

In addition to suffering habitat loss and degradation, the native land snails of Sāmoa and other islands in the Pacific are under assault by introduced predators. One of these predators is the rat. The other is another nonnative land snail, the rosy wolfsnail (*Euglandina rosea*), whose dietary preference is smaller snails.

This is a story about snails gone awry: The giant African snail (*Achatina fulica*) was introduced to American Sāmoa sometime prior to 1977, probably by Asian commercial fishermen as a source of food. The population of this herbivorous snail quickly grew and spread, eating all kinds of vegetation and causing damage to agricultural crops. Not only is the snail an agricultural pest, but it is also a carrier of the rat lungworm, a parasite that causes eosinophilic meningoencephalitis in humans. Efforts to eradicate the snails by collecting and killing them were unsuccessful. During an eight-month period during 1980, more than five million snails were collected, but the population did not decline.

In a second effort to get rid of these vegetarian giants, two smaller carnivorous snails, the rosy wolfsnail (*Euglandina rosea*) and the carnivorous snail (*Gonaxis kibweziensis*), were introduced in hopes they would eat their giant African relative, which can grow to over 20 cm (8 in.) in body length. Catastrophically, yet predictably, the introduced carnivorous rosy wolfsnail prefers to eat smaller native snails instead of the African giant. The rosy wolfsnail is now the culprit behind the endangerment—and extinction—of native snails.

As mentioned, many of Sāmoa's snails are endemic to the archipelago, existing nowhere else in the world. Further, a large proportion of these species is known from only single islands. Unfortunately, of the 15 species endemic to American Sāmoa, only six species were found alive during surveys in 1993. For example, only one individual of *Sāmoana thurstoni*, endemic to the island of Ofu, was found alive; surveys in 1998 located only a dozen. Because only the shells of dead *Sāmoana abbreviata*, endemic to Tutuila, were found during 1993 surveys, it was feared that this species had become extinct. Although snail surveys in 1998 located several individuals of this species, their numbers remain precariously low, and *S. abbreviata* may still be in jeopardy of extinction. Sadly, the Mount Matafao snail (*Diastole matafaoi*) is extinct (IUCN Red List: EX). Further, rat predation on land snails is a source of mortality all too frequently observed (in shells missing a large piece, indicative of a bite). As of yet, the introduced carnivorous snails have not found their way to the western islands of Sāmoa.

The extinction of native terrestrial plants and animals as a result of introduced species and habitat loss is an all too common occurrence on Pacific islands. How many times will we make the same mistake by introducing a nonnative species in hopes of solving some problem, a problem we probably created in the first place?

Sāmoana conica (endemic to the archipelago; left-handed), IUCN Red List: EN

Sāmoana thurstoni (endemic to the island of Ofu), IUCN Red List: EN

Sāmoana abbreviata, IUCN Red List: CR

This species is almost identical to *Sāmoana conica* but is right-handed.

Succinea crocata (endemic to the archipelago)
This tree snail has a thin, almost transparent shell, with much of the golden-brown coloration radiating through the shell from the snail's body inside.

Eua expansa (endemic to the western islands of Sāmoa)

Eua zebrina (endemic to the island of Tutuila), IUCN Red List: EN
The hanging chandeliers in the lobby of the Rainmaker Hotel on Tutuila are decorated with an estimated 10,000 partulid snail shells, mostly the precariously rare *Eua zebrina*.

Trochomorpha apia (endemic to the archipelago)

Ostodes strigatus (endemic to the island of Tutuila)

Pythia scarabaeus (occurs throughout the Pacific)

Nonnative (Introduced) Species

Giant African snail, *Achatina fulica*

Bradybaena similaris

Rosy wolfsnail, *Euglandina rosea*

Laevicaulis alte
gau, **gaupapa**, **matefānau**

Subulina octona
This species is common in forest habitats and around
dwellings.

INSECTS AND ARTHROPODS
(CLASS INSECTA AND ARTHROPODA)

There are 2,775 species of insects and other arthropods recorded for the Sāmoan Archipelago, too many to be covered here. Nevertheless, I have included butterflies and a few additional noteworthy or widespread residents in other insect groups. For a complete list of all insects and related arthropods recorded for the archipelago and its individual islands, see Kami and Miller (1998).

Millipedes and Centipedes

Millipedes (**taetuli**) and centipedes (**atualoa**) are both members of the phylum Arthropoda, which contains approximately 838,000 different species. Arthropods are covered with a rather hard, jointed outer skeleton. Millipedes and centipedes have many more legs than their six-legged cousins the insects (class Insecta).

Millipedes are members of class Myriapoda. The word "millipede" is Latin for "a thousand feet." Nine species of millipedes have been recorded from the archipelago. Millipedes are rounded and tubular in shape, with two pairs of legs on each body segment. They often roll up into a tight, protective spiral when touched (which makes them easier to pick up and toss out the door).

Centipedes are members of the class Chilopoda. The word "centipede" is Latin for "a hundred feet." Six species of centipedes have been recorded for the archipelago. Centipedes have a more flattened shape than millipedes, and a tough exterior that is difficult to crush. Although only a single pair of legs occurs on each well-defined body segment, centipedes move with incredible speed. When you see a centipede, the term "creepy-crawly" immediately comes to mind. Centipedes differ from millipedes in another important way: millipedes are rather harmless vegetarians, while centipedes are carnivores. The centipede's front pair of legs has evolved into poison pincers capable of inflicting a terrible bite. If bitten, a person may experience intense localized pain, swelling, vomiting, dizziness, headache, and even an irregular pulse. Some tropical centipedes are known to reach 30 cm (12 in.) in length, but the largest species in the archipelago (*Scolopendra subspinipes*) is generally about 15 cm (6 in.), with twenty pairs of walking legs. After being bitten in the leg by a centipede during the middle of the night, I find this species no less intimidating than its larger relatives and, several years later, still have recurring "centipede dreams."

Cockroaches

Cockroaches (**mogamoga**) are members of class Insecta, order Blattaria. They are the most primitive of winged insects, appearing in the fossil record over 300 million years ago. Although they can fly, most seem to prefer to scurry along the ground, floors, walls, and everything in between. Most species are nocturnal and prefer to hide in dark places during the day. Special sensory structures on their abdomens sense minute air movements, thus making cockroaches difficult to catch or squish. The brown, rectangular egg case contains about a dozen compartments.

The American cockroach (*Periplaneta americana*) takes about 12 months to develop into an adult, molting many times during the year until reaching its full size of about 5 cm (2 in.). The adults may live for another year. While cockroaches eat all kinds of food, they are also an important part of the food web themselves and often end up as a dinner item for others species, such as geckos. Twenty-three species of cockroaches are recorded for the archipelago.

Beetles

Beetles are one of the most numerous animal groups on the planet with an estimated 3 million species worldwide, perhaps more. To date, 536 beetles species are recorded for the archipelago, though there are probably many more left to discover.

Brick red millipede, *Trigoniulus corallinus*
taetuli

Common large brown millipede, *Leptogoniulus sorornus*
taetuli

Small brown millipede, *Pseudospirobolellus avernus*
taetuli

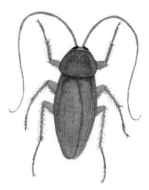

American cockroach, *Periplaneta americana*
mogamoga

Centipede, *Scolopendra subspinipes*
atualoa

Giant crab spider, *Heteropoda venatoria*
Also known as the "banana spider" (since it may hide in
banana bunches), this is an important spider that feeds on
cockroaches and other domestic pests—perhaps even centipedes. It does not make webs, but rather uses speed to capture prey.

This spider's very flat body enables it to fit into small crevices and spaces. It is well adapted to life in, or near, human habitation, so it may often be seen in houses, sheds, under or between boards, etc.

Though its large size (up to 12 cm, or 5 in.) may be disarming, it is neither dangerous nor poisonous to humans. However, if handled, it could inflict a painful bite. It is presumed to be native to Asia, but now occurs worldwide, particularly in tropical or subtropical areas.

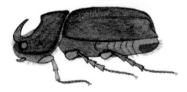

Rhinoceros beetle, *Orcytes rhinoceros*

manu'ainiu

The rhinoceros beetle is the most destructive pest of the coconut palm (*Cocos nucifera*). It tunnels into the tender portions of the crown of the palm, where it drinks the sap, thereby negatively affecting the number of coconuts produced by the tree. It breeds in dead coconut trees and other dead wood, as well as in other decaying matter such as piles of rubbish, sawdust, manure, etc.

Butterflies

Most butterflies inhabiting the archipelago are Indo-Malayan (from the region extending from East Africa to Malaysia) in origin and are widespread throughout Polynesia. Like birds, some species of butterflies are highly mobile and, aided by winds, are capable of dispersing over long distances, including oceans.

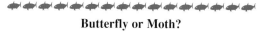

Butterfly or Moth?

The taxonomic separation of moths and butterflies is vague, and not all species exhibit the exemplifying characteristics. Generally speaking, butterflies are diurnal (active during the day), have clubbed antennae, and are often brightly colored. Conversely, moths are generally nocturnal (active at night), usually have tapered or featherlike antennae, and are dull colored.

The butterfly wing consists of many delicate scales that overlap like tiles on a roof. Butterflies have no jaws and always take their food in liquid form using a modified tongue, or proboscis. The butterfly's proboscis is a long, hollow tube (like a straw) that is coiled like a roll of tape. When the butterfly is feeding, the proboscis unrolls to probe deep into flowers in search of nectar.

The butterfly life cycle is extraordinary. The butterfly undergoes a complete change, or "metamorphosis," during the course of its development. This miracle of nature transforms a frequently ugly, prickly, and bizarre-looking caterpillar into an exquisite butterfly. The life cycle includes several stages before the adult butterfly emerges, from egg, to larva (caterpillar), to pupa (chrysalis), and finally, adult.

Female butterflies lay their eggs on, or very near, the food plant upon which the caterpillar will feed. Frequently, the food plant must be a specific species or "host plant" in order for the caterpillar to survive. The eggs are usually yellow or green and laid on the undersides of leaves so they receive some protection from predators, rain, and sun.

From the eggs hatch caterpillars. The caterpillar stage is primarily a feeding stage, during which large quantities of vegetation are consumed. The next stage in the life cycle is the pupal stage. The caterpillar's skin shrivels and splits to expose the pupa. During the pupal, or "chrysalis" stage, much of the caterpillar's body metamorphoses (transforms) into the components of the adult butterfly (with wings). The pupa is immobile and does not eat or drink. While the duration of the life cycle varies in different species, the pupa eventually splits and the butterfly emerges. Upon emergence, the wings are soft and crumpled, but they soon expand by "pumping up" with blood. After they dry and harden, the butterfly takes wing.

Many butterflies fly only during times of bright sunshine. Further, some species have special habitat requirements or extremely specialized food requirements. For example, butterflies and other insects of the coastal areas of the archipelago are not usually found at higher elevations unless clearings, deforestation from logging, and roads have opened up the dense montane forest.

Sadly, butterflies are threatened by habitat deterioration and destruction. Even if the adults could adapt to new conditions, the caterpillars probably would not survive because of their explicit need for specific food plants.

Due to unresolved taxonomy, various authors report differing numbers of butterfly species for the archipelago. Classification remains controversial, and many of the species and subspecies names applied to Sāmoan insects vary within the literature.

For a complete list of all butterflies recorded for the archipelago, see Kami and Miller (1998). In their Checklist of Sāmoan Insects, they list 33 species. One additional species (*Doleschallia tongana*), has recently been recorded for the island of Tutuila. The Checklist of Sāmoan Insects lists three additional subspecies. Many subspecies, some of which may be endemic, await taxonomic clarification. In addition, confusing variation in wing pattern can, and does, occur between individuals of the same species. Five of the species have been recorded for remote Swains Island, but none have been recorded for Rose Atoll. Only one species, a swallowtail (*Papilio godeffroyi*), is endemic to the archipelago. Of the 15 butterfly families in the world, 5 families are represented in the archipelago: Hesperiidae (2 species), Lycaenidae (10 species), Nymphalidae (16 species), Papilionidae (1 species), and Pieridae (5 species).

Family Nyphalidae:

Monarch, *Danaus plexippus*

Euploea lewinii bourkii

male   female

Hypolimnas bolina pallescens

This is the most conspicuous butterfly in American Sāmoa; the sexes differ dramatically.

Hypolimnas antilope lutescens

Junonia villida villida
This butterfly is common in coastal areas.

Melantis leda

Tirumala hamata
This butterfly is rare in American Sāmoa, but abundant in some areas of the western islands of Sāmoa.

Vagrans bowdenia

Family Lycanidae (a family filled with similar, hard-to-identify confusing little butterflies, many of them blue):

Zizina labradus

Family Papilionidae:

Sāmoan swallowtail, *Papilio godeffroyi* (endemic to the archipelago)

Family Pieridae:

Appias sp.

THE PROTECTED RESERVES OF SĀMOA

AMERICAN SĀMOA
National Park of American Sāmoa

The National Park of American Sāmoa ranks with better-known U.S. national parks such as Yellowstone and the Grand Canyon, but because it was established just recently, few people know about it. The park is in its earliest stages of development, but what it lacks in visitor facilities it makes up for in South Pacific splendor.

The National Park of American Sāmoa was authorized by the United States Congress in 1988 to preserve spectacular old-world (paleotropical) rain forest, stunning Indo-Pacific coral reef, and the 3,000-year-old Sāmoan culture known as **fa'asāmoa**. This is the only U.S. national park in which the United States federal government leases the land from local villages instead of owning and managing the land. In 1993, the park was officially established when a 50-year lease was signed between the U.S. government, the American Sāmoa Government, and village chiefs (**matai**).

The 4,250-ha (10,500-a.) park encompasses portions of the three rugged, volcanic islands of Tutuila, Ofu, and Ta'u. Over 1,000 ha (2,500 a.) of the park are underwater. Habitats vary in topography from reef, beach, and rocky coast all the way up to cliffs soaring 945 m (3,100 ft.) above the ocean. The Tutuila and Ta'u units encompass native rain forest, with small patches of agricultural plantations. The Ofu unit encompasses a coral reef and adjacent beach, providing habitat for a multitude of coral reef fish and coastal strand vegetation. The steep rock cliffs of the park provide habitat for numerous species of seabirds, and the rugged mountainsides to forest birds, flying fox fruit bats, ferns, orchids, and numerous other animals and plants. The offshore coral reef and rocky coastlines provide habitat for the 1,000 or so species of coral reef and pelagic fishes that are known to occur in the archipelago. (The Caribbean, by contrast, has only about 300 species of fishes.)

The 2,025-ha (5,000+-a.) Ta'u unit is the largest unit of the park, and encompasses mostly undisturbed coastal, lowland, montane, and cloud rain forest communities. Laufuti Falls, on the south coast, plunges over 300 m (1000+ ft.) to the sea.

The 607-ha (1500-a.) Ofu unit, only 14.5 km (9 mi.) across open ocean from Ta'u, primarily encompasses an exceptionally beautiful coral reef and lagoon bordered by a white sand beach. Here, hundreds of species of reef fishes, corals, and marine creatures can be observed in the crystal-clear water. This beach, with its view of mist-shrouded rain forest pinnacles and beautifully sculptured volcanic boulders, is one of the most exquisite in the entire South Pacific. This unit offers the best, most accessible snorkeling in the park.

The 1,417-ha (3,500+-a.) Tutuila unit, like Ta'u, encompasses primarily undisturbed coastal, lowland, and montane rain forest, rising from the rocky north coast to the top of Mt. Alava, rising 491 m (1,610 ft.) above Pago Pago harbor. The Amalau Valley, nestled on the north coast, offers a wonderful roadside view of many species of birds and both species of flying fox fruit bat. Just west is the rock formation known locally as the Pola (or Cockscomb), rising over 128 m (420 ft.) out of the sea. These rocks, which resemble a dinosaur tail jutting 1,060 m (3,500 ft.) out into the sea, are a favorite haunt of Tutuila's seabirds.

Mt. Alava affords wonderful views of the Tutuila unit of the park, and nearly all of the island. On a clear day, the neighboring western islands of Sāmoa and the islands of Manu'a are visible. The Mt. Alava trailhead begins at Fagasa Pass on the crest of the pass between Pago Pago and Fagasa villages. This hiking trail, which is really a dirt road originally built to provide access to television transmitters on the top of Mt. Alava, was recently improved, but may be pitted with small, muddy puddles. It is an approximately three-hour moderate hike to the top. Take plenty of water and sunscreen.

For more information on the various park units, hiking trails, and current conditions, contact park headquarters on Tutuila.

Fagatele Bay National Marine Sanctuary

Fagatele Bay National Marine Sanctuary is one of over a dozen national marine sanctuaries in the United States. It was designated a marine sanctuary by Congress in 1986 to protect a 0.65-sq.-km (0.25-sq.-mi.) coral reef on the south shore of Tutuila. The tiny jewel of Fagatele Bay is formed by the crater of an extinct volcano, with one side now open to the sea. The vertical cliffs surrounding the bay are covered in undisturbed rain forest vegetation; other areas within the crater are plantations. A variety of seabirds nest and forage in the sanctuary, and the native rain forest vegetation provides habitat and safe roosting areas for the white-necked flying fox fruit bat (*Pteropus tonganus*). Both the green and the hawksbill sea turtles have been seen in or near Fagatele Bay.

Nearly 200 species of corals are recovering from the island-wide invasion of crown-of-thorns starfish (*Acanthaster planci*) in the late 1970s, which destroyed 90% of the corals. The corals here and elsewhere around the archipelago were subsequently devastated by the hurricanes of 1990 and 1991 and a massive coral bleaching event in 1994. Most recently, the reef flats were affected by El Niño, which caused exceptionally low tides throughout the western Pacific. Corals and many related animals were killed on some of the exposed reef areas.

The national marine sanctuary program serves as the

trustee for the United States' major system of marine protected areas. Sanctuaries conserve, protect, and enhance the biodiversity, ecological integrity, and cultural legacy of these ecosystems. The American Sāmoa Department of Commerce administers the Sanctuary under a cooperative agreement with the National Oceanic Atmospheric Administration (NOAA).

The sanctuary serves as a natural classroom and laboratory for residents of Sāmoa. Activities include school and village educational outreach programs, enforcement of regulations, and a long-term marine research and monitoring program. Fagatele Bay contains recreational opportunities for diving, snorkeling, limited sport and commercial fishing, swimming, and picnicking. Call the sanctuary office for information on visiting.

Vaoto Marine Reserve

The Vaoto Marine Reserve is located on the island of Ofu, directly adjacent to the Ofu Unit of the National Park of American Sāmoa. This approximately 16-ha (40-a.) territorial reserve is located directly seaward of the airport landing strip and Vaoto Lodge. It was preserved because of its proximity to the airport and ease of access for visitors and tourists to a beautiful coral reef and sandy beach. The Vaoto Marine Reserve also contains an unusually high concentration of the rare blue coral (*Heliopora coerulea*) (see Marine Invertebrates chapter).

Nu'uuli and Leone Pala

The Nu'uuli Pala is the largest mangrove swamp on Tutuila. The wetland (excluding the open water of the lagoon) covers 50 ha (123 a.) of mangrove forest and mangrove swamp. The majority of the **pala** (swamp) is covered with Oriental mangrove and red mangrove. Several areas of other freshwater marsh vegetation are interspersed with the mangroves. A small area of saltwater marsh borders the end of Coconut Point.

This area is extremely important as fish and wildlife habitat. It is also important for subsistence use, and provides recreational opportunities such as fishing and canoeing. Mangrove habitats provide an important crab fishery as well. Unfortunately, most of the freshwater marsh and mangrove forest has been degraded by clearing, filling, dumping of household and industrial trash, and proximity to piggeries.

Farther west in the village of Leone is the 9-ha (21-a.) Leone Pala. Leone Pala is confronting the same environmental challenges facing Nu'uuli Pala, but since it is much smaller than the Nu'uuli Pala, it is therefore perhaps more vulnerable and sensitive to disturbance. The American Sāmoa Environmental Protection Agency and Coastal Zone Management Program are working to protect the remaining wetland habitats of Sāmoa. Both Nu'uuli Pala and Leone Pala have been designated Special Management Areas under the American Sāmoa Coastal Management Act of 1990.

Rose Atoll National Wildlife Refuge

Rose Atoll is a remote coral reef in the territory of American Sāmoa, situated more than 150 km from the nearest island to the west, Ta'u. It was designated a United States National Wildlife Refuge in 1973 "for the conservation, management, and protection of its unique and valuable fish and wildlife resources." It is one of the smallest coral atolls in the world (654 ha, 1,613 a. [1,593 submerged, 20 emergent]). A fringing circular reef approximately 1.6 km (1 mi.) in diameter surrounds a central lagoon (14 m deep). The reef supports two emergent islands, Rose Island (7.3 ha) and Sand Island (0.8 ha).

The habitat is unique in the archipelago. The reef flats of the atoll are dominated by crustose pink coralline algae. Its current name apparently refers not to the color of the coralline algae but to the wife of an exploring sea captain who visited the atoll in 1819. The atoll was first described by European explorers almost 100 years earlier and had several other names prior to being called Rose Atoll.

The deep, clear waters of the lagoon are almost entirely enclosed by the reef, except for a narrow, deep channel open to the sea. The floor of the lagoon is mostly sand, with several patch reefs and coral pinnacles. The pinnacles rise the depth of the lagoon and are covered with algae, coral, and giant clams. Rose Island contains a large grove of **pu'a vai** trees (*Pisonia grandis*), and only six other species of plants. An unusually high concentration of salt in the soils may be the reason for the low plant diversity.

On the other hand, for its size, Rose Atoll is a diverse and extremely important site for federally protected seabirds and nesting sea turtles. It supports 97% of the total number of nesting seabirds in American Sāmoa and is home to 20 species of seabirds; 12 species nest there, and 8 other species have been observed on the islets. As of 1998, seabirds nest in numbers totaling approximately 273,000 individuals, and approximately 270,000 of these are sooty terns (*Sterna fuscata*). In addition to sooty terns, the nesting colonies include greater frigatebird (*Fregata minor*) and lesser frigatebird (*Fregata ariel*), red-footed booby (*Sula sula*), brown booby (*Sula leucogaster*), masked booby (*Sula dactylatra*), red-tailed tropicbird (*Phaethon rubricauda*), white (fairy) tern (*Gygis alba*), black-naped tern (*Sterna sumatrana*), blue-gray noddy (*Procelsterna cerulea*), and black noddy (*Anous minutus*). Both the green sea turtle (*Chelonia mydas*) and hawksbill sea turtle (*Eretmochelys imbricata*) also nest at the atoll. The green sea turtles are known to range great distances throughout Polynesia before coming to Rose and Sand islets to nest (See Reptiles and Amphibians chapter).

The reefs provide an important refuge for numerous

marine species, including the giant clam (*Tridacna maxima*), which has been overharvested in the rest of the archipelago. Over 200 species of fishes have been recorded from the atoll. The fish community is distinctly different from those of waters elsewhere in the archipelago, including fewer herbivorous species (e.g., damselfishes and parrotfishes), but more carnivorous species (unicornfishes and snappers). Blacktip reef sharks are frequently observed, often with dorsal fins emergent as they cruise up and down the shallow areas of the lagoon. Groups of wintering humpback whales have been seen just offshore, filling the ocean with their exquisite, haunting songs.

Several nonnative species have also made their way to this remote location. The Polynesian rat (*Rattus exulans*), accidentally introduced to the atoll, was responsible for the decimation of much of the atoll's nesting seabird population. The American Sāmoa Department of Marine and Wildlife Resources and the U.S. Fish and Wildlife Service began a rat eradication project in 1990, and current surveys indicate that all rats have now been removed from the islets. Seabirds can now nest without harm to their eggs or themselves. Further, scientists have also recently been successful at eradicating nonnative plant species from the islets. Great care is taken by scientists visiting the atoll not to accidentally reintroduce nonnative plants by inadvertently dropping seeds from personal belongings and effects used in other island locations throughout the Pacific or elsewhere. Every item of clothing, and all other effects such as backpacks or tents, worn or brought onto the island must be brand new or previously used only at Rose Atoll.

Although the atoll is now uninhabited by people, there is reference to a few inhabitants of German and Sāmoan descent who attempted to establish a fishing station during the 1860s. Today, access is provided only by special, scientific permit. These restrictions exist to protect the native wildlife species using Rose Atoll and its fragile environments.

The primary mission of the U.S. Fish and Wildlife Service is to acquire, protect, and manage unique ecosystems necessary to sustain fish and wildlife such as migratory birds, resident species, and endangered species. The refuge is jointly managed by the U.S. Fish and Wildlife Service and American Sāmoa Government Department of Marine and Wildlife Resources.

〜〜〜〜〜〜〜〜〜〜〜〜〜〜〜〜〜〜

While Rose Atoll was once one of the most pristine reefs in the world, in October of 1993 a large Taiwanese longline fishing boat ran aground, spilling over 370,000 liters (100,000 gal.) of diesel fuel, almost 1,900 liters (500 gal.) of lubricating oil, and over 1,100 kg (2,500 lbs.) of ammonia onto the reef. The accident caused widespread injury to the marine life at the atoll and physically devastated and degraded important coral reef habitat. Chemical contamination persisted for at least two years after the spill, changing the composition of the marine flora and fauna. Tons of metal and other debris were strewn across areas of the reef. Although salvage operations removed a huge portion, much wreckage remains. The movement of this remaining wreckage by waves and currents has continued to break and damage corals, clams, and other marine life. In addition, an unnaturally large amount of blue-green algae is thriving on the reef in the area of impact.

Scientists continue to monitor the impacts of this spill and the recovery and deterioration of this special reef ecosystem.

〜〜〜〜〜〜〜〜〜〜〜〜〜〜〜〜〜〜

National Natural Landmarks

American Sāmoa is fortunate to have seven areas designated as National Natural Landmarks on the island of Tutuila. This program is also administered by the U.S. National Park Service. National Natural Landmarks are areas that contain unique ecological or geological features. Diamond Head Crater in Hawaiʻi is a well-known example of a National Natural Landmark. The National Natural Landmarks Program was established by the National Park Service to identify and encourage local preservation of unique sites, but there are no federal laws to protect these sites. In addition, all sites on Tutuila are not part of the National Park of American Sāmoa, with the exception of Vaiava Strait.

The mission of this program is to (1) encourage the preservation of sites illustrating the geological and ecological character of the United States; (2) enhance the scientific and educational value of such sites; (3) strengthen public appreciation of natural history; and (4) foster a greater concern for the conservation of our nation's natural heritage. Following are brief descriptions of these seven stunningly beautiful areas.

ʻAunuʻu Island

ʻAunuʻu is located off the eastern coast of Tutuila, and is less than 1.6 sq. km (1 sq. mi.) in area. About half the island is a cultivated plain on which sit the village of ʻAunuʻu and associated plantations. The island is rimmed here and there by beaches composed of coral rubble and sand and sea cliffs. The eastern half of the island is rimmed by a geologically recent volcanic cone. The highest point on the crater rim is 95 m (310 ft.) above sea level. Within the crater is Faʻimulivai Marsh, which includes a freshwater pond. This crater is one of the few places where one can observe the more recent episodes of volcanism in American Sāmoa.

The eruptions of 'Aunu'u occurred at approximately the same time as the volcanism of the Leala shoreline (below), while the eruptions that formed the volcanic plugs of Rainmaker Mountain, Matafao Peak, and the ridges near Vaiava Strait are much older.

This island has the only quicksand area and freshwater lake in American Sāmoa. Just north of the village is Pala Lake, a beautiful but potentially dangerous area of quicksand. On the far eastern edge of the crater is Ma'ama'a Cove, a beautiful but deadly juxtaposition of eroded sea cliffs, ocean, and waves. View the water of the cove from a distance; wave action is extremely unpredictable, and large and unexpected waves appearing out of nowhere wash away everything in their path.

Cape Taputapu (Forbidden Beach)

Cape Taputapu ("taboo-taboo") is a fantastic fairyland of shoreline and offshore volcanic rocks and blowholes sculpted by heavy wave action. Huge sculptures of erosion-resistant volcanic rock dot the water offshore. One of these islets has been identified by scientists as a volcanic vent through which lava poured during the major episode of volcanism that formed Tutuila Island. The land and water designated within this site total 69 ha (170 a.). Cape Taputapu is the westernmost point of Tutuila, located just beyond the village of 'Amanave.

Fogāma'a Crater

Fogāma'a Crater is a 196-ha (485-a.) site located on the southwest shoreline of Tutuila, in the village of Futiga. This crater lies just inland from Larsen Bay, which encompasses two smaller coves, Fogāma'a and Fagalua. A view of its distinctively craterlike appearance is best appreciated from offshore. Fogāma'a Crater illustrates the most recent episode of volcanism in American Sāmoa. During the time since eruptions from this crater occurred, it has been breached by erosion on its seaward side. It is now surrounded by coconut plantation.

Leala Shoreline (Sliding Rock)

The Leala Shoreline is a rugged and spectacular exposure of basaltic rock flows located just south of the village of Taputimu on the southwest coast of Tutuila. Beginning near Vailoatai Village on the south shore, it runs east to Fagatele Bay National Marine Sanctuary. The landmark comprises 14 ha (35 a.) of this shoreline. The rock above the upper limits of erosion by wave action is covered with dense coastal vegetation. The gently sloping rock shelf is known locally as "sliding rock," as much of the area is covered with algae and is extremely slippery when wet. The basalt in the intertidal zone is pitted with pools, creating unique rocky-intertidal habitat. The marine life of these tidal pools is replenished by breaking waves and rising tides. One of the pools is approximately 30 m long and more than 3 m deep.

This landmark site is both ecologically and geologically spectacular. If you are visiting the site, be careful, because the algae-covered rock is extremely slippery and wave action is unpredictable: I am one of many unwary visitors who have been caught off guard by a phantom wave, escaping with only a few bruises. Other unlucky visitors have been washed into the sea by these rogue waves.

Matafao Peak

Matafao Peak is the highest peak on Tutuila, rising above the island to an elevation of 652 m (2,142 ft.). Along with Rainmaker Mountain (Mt. Pioa), Matafao is one of the five great masses of volcanic rock extruded as molten magma during the major episodes of volcanism that created Tutuila Island. It is one of the most impressive topographic features on the island. The site comprises 71 ha (175 a.), the area above the 365-m (1,200-ft.) contour line. An unimproved, steep trail runs up the eastern side of the mountain. The trailhead begins at the top of Fagasa Pass, just across from the entrance to the Mt. Alava trail; it is best to find someone who knows the area to show you. The hiking trail is difficult and will take most of a day. Take plenty of water and sunscreen.

Rainmaker Mountain

Rainmaker Mountain, or Mt. Pioa, is one of several giant volcanic plugs associated with the earliest major episodes of volcanism that created Tutuila Island. Steeped in Sāmoan lore and legend, it dominates the scene from almost any point in the Pago Pago Harbor area and comprises three peaks: North Pioa, South Pioa, and Sinapioa. The peaks range in elevation from 480–520 m (1,619–1,718 ft.). The 69-ha (170-a.) area designated as a landmark occurs above the 245-m (800-ft.) contour line.

Pristine montane scrub vegetation, home to several endemic plants that grow only here and on Matafao Peak, cap the peaks. The top of Afono Pass, which switchbacks steeply up from the village of Aua, is a good place to view flying fox fruit bats and white-tailed tropicbirds.

Vaiava Strait (and Pola Island)

Vaiava Strait is located on the north coast of Tutuila, on the northwestern side of the village of Vatia. Although designated a 101-ha (250-a.) National Natural Landmark site, it is also a small part of the lands and waters comprised by the National Park of American Sāmoa. Here, erosion by the sea has sculpted deep and scenically spectacular cliffs and sea arches in the rocks of a huge volcanic plug, locally known as Pola. Its dramatic 128-m (420-ft.) cliffs and rock top are an important resting and nesting area for many seabird species.

WESTERN ISLANDS OF SĀMOA
Falealupo Preserve

Falealupo Preserve is located on the extreme western end of the island of Savai'i. This preserve came to be because the village of Falealupo needed money to build a new school. When there is a lack of regular income and available funds for village improvements, one way to secure such funds is for the village to lease its forested lands for logging. The village leaders decided instead to work with private donors interested in conservation to prevent the loss of their beautiful, undisturbed lowland rain forest to the loggers' bulldozers. In 1989, chiefs from the village of Falealupo signed a covenant to preserve their forest for a period of at least 50 years. In exchange, money for a new school was provided. The covenant provides for the continued village use of the forest for cultural and agricultural purposes, including collecting medicinal plants and cutting timber for kava bowls and canoe construction.

As you enter the preserve, stop by the new Rain Forest Canopy Walkway, which was constructed in 1997 to offer visitors a bird's-eye view of the forest and its inhabitants. Visitors climb up stairs that wind around the trunk of a sinuous banyan tree, then cross a dangling walkway to another tree. This walkway and series of platforms culminate in a spectacular viewing platform 30-m (100-ft.) high in the canopy of the trees.

Tafua Peninsula Rain Forest Preserve

Three villages have communal land composing the Tafua Peninsula: Tafua, Faala, and Salelologa. Though the area was first identified as a potential national park site in 1975 because it was the largest contiguous area of undisturbed lowland rain forest remaining on Savai'i, initial preservation efforts were unsuccessful, and it was not until the early 1990s that the area was protected. Just prior to its preservation, the village of Tafua was faced with a problem similar to that of Falealupo. The villagers desperately needed funds to build a school and were receiving pressure from logging companies to harvest their forest. Similarly, Faala Village wanted to build a road to Aganoa Beach. Instead of selling their forest trees to the logging company, the villages that owned the forest worked with the O le Siosiomaga Society Inc. and the Swedish Nature Society to raise the needed funds for the school and the road. The village of Tafua also enlisted the help of another local conservation group, Faasao Savai'i Inc. In return, the three villages set aside forested lands that now compose the 4,450-ha (11,000-a.) Tafua Peninsula Rain Forest Preserve. this agreement, however, may be unraveling.

The peninsula is partly surrounded by barrier reef. Other coastal edges are lava cliffs, rising 3–16 m (10–40 ft.) above the sea. The preserve is biologically diverse both in flora and fauna. Bird species abound; 25 species have been recorded, including the endemic tooth-billed pigeon (*Didunculus strigirostris*, **manumea**) and rare many-colored fruit dove (*Ptilinopus perousii*, **manumā** [female], **manulua** [male]). The forest provides medicinal plants for use by traditional healers, and wood for sanctioned construction such as canoes. Aganoa Beach in Faaala, and Niutala Beach in Salelologa are nesting sites for hawksbill sea turtles (*Eretmochelys imbricata*), and humpback whales (*Megaptera novaengliae*) may be observed offshore from July through November. A dramatic, steeply-walled crater located at the crest of Mt. Tafua is often a good bat-viewing location, and a hiking trail leads up to the crater rim. The preserve is located on the southeastern side of Savai'i; make arrangements in the village of Tafua to enter the preserve.

O le Pupu Pu'e National Park

Created in 1978, O le Pupu Pu'e National Park (30 sq.

km) is located on the southern side of the island of 'Upolu. It comprises a wedge of island habitats extending from the cloud forest of the island's backbone and mountains—Mt. Fito, 3,609 ft. (1,100 m), and Mt. Le Pu'e, 2,756 ft. (840 m)—to the lava cliffs of the coastline. Because of its span, it preserves a diverse array of coastal, lowland, and montane plant species. Although recent hurricanes have damaged the forest, it is still home to many of Sāmoa's animal species.

Much of the park's present area was formed just 3,000 years ago when Mt. Fito last erupted. This spread a massive flow of lava over its lower elevations. These basalt rocks have just begun to weather, and there is little soil present. The higher elevations comprise forests bisected by deep valleys and several volcanic craters. Mt. Le Pu'e has a lake in its crater.

There is an easy trail along the cliffs of the coast, but it is rocky and may be overgrown. Be careful. The trail proceeds over lava rocks through forests of pandanus and other coastal vegetation. There are nice coastal viewing areas along the trail, including sea arches and blowholes. Check trail conditions with a ranger before visiting this and other areas of the park. The Division of Environment and Conservation has an office in downtown Apia.

The Pe'ape'a Cave and trails are for the more adventurous hiker. The cave is actually the inside of a lava tube. As the lava flowed, its outside edges, in contact with the cooler air, hardened into a crust. In contrast, the interior remained molten and continued to flow. Eventually, the flow ceased, leaving the lava tube crust and cave behind. This trail begins in the visitor center area, but it would be best to check trail conditions with a ranger or go with a guide knowledgeable about the area. The trail may be overgrown, and trails made by feral pigs may be confusing. The cave is allegedly named after the **pe'ape'a** (white-rumped swiftlet), which nests in small numbers inside. The small sheath-tailed bat (**tagiti**) is often confused with the **peʻapeʻa** and perhaps roosted here at one time (see Mammals chapter). The entrance to the cave itself is a 10-m (30-ft.) drop that can be extremely slippery and dangerous.

O le Pupu Puʻe National Park is managed by the Department of Lands, Surveys, and Environment. Call for additional information and trail conditions.

Togitogiga Recreation Reserve

The Togitogiga Recreation Reserve is located within O le Pupu Puʻe National Park, fairly close to the main road. Unfortunately, the visitor center was destroyed during the hurricanes of 1990 and 1991 and has not been rebuilt. A small yet beautiful "swimming pool," created by the main waterfall of the Mataloa River, flows through the reserve. Several other swimming areas are located along the river. Care should be taken while swimming, as currents can be swift, and hidden logs, rocks, and boulders abound. This reserve, also managed by the Department of Lands, Surveys, and Environment, is easily accessible by vehicle along the south coast of ʻUpolu, just east of Cross Island Road.

Palolo Deep Marine Reserve

The Palolo Deep Marine Reserve, established in 1974, is a fringing reef encompassing a "deep" (a hole) in the reef where numerous marine species are easily observed (108 species of fishes have been recorded). The most interesting feature of the reserve is the deep itself, a steep-sided lagoon occurring in the middle of the reef flat. Such holes are common in the reefs of the western islands of Sāmoa. The reserve itself comprises 137.5 ha (56 a.); the hole is about 200 m in diameter by 10 m deep. While recent hurricanes have damaged the coral, it is recovering.

Access is moderately easy, as the deep itself is located approximately 100 m (330 ft.) from the shoreline. At low tide, the water over the reef flat leading out to the deep is quite shallow, making it necessary to walk over the coral rubble to reach the deep; to avoid damaging the coral, go when water levels are higher and you can snorkel out to the deep. Periods of high tide or rough ocean can occasionally make it difficult to reach the deep; consult local tides and the marine forecast.

Palolo Deep Marine Reserve is located just off Beach Road, behind the main wharf and within walking distance of the center of Apia. Masks, snorkels, and fins can be rented, and there is a small charge to visit the site. Picnic **fale** and restrooms are available. Palolo Deep is privately managed with assistance from the Department of Lands, Surveys, and Environment.

Mount Vaea Reserve: Mount Vaea Scenic Reserve, Vailima Botanical Garden, and Robert Louis Stevenson Memorial Reserve

The Mount Vaea Reserve encompasses three sites: the Mount Vaea Scenic Reserve, Vailima Botanical Garden, and Robert Louis Stevenson Memorial Reserve. Directly adjacent to these Mount Vaea Reserve sites is the Robert Louis Stevenson Museum.

The Mount Vaea Scenic Reserve (62 ha, 128 a.) covers the eastern-facing slopes of Mount Vaea. Mount Vaea is the prominent mountain overlooking Apia. While much of the forest has been disturbed and modified by past activities, including recent hurricanes, beautiful rain forest with a variety of wildlife still persists. The "Road of Loving Hearts" leads visitors through the reserve to the Robert Louis Stevenson Memorial Reserve and a beautiful view of Apia far below. Two trails lead to the Memorial Reserve. Though both are moderately strenuous, the long trail is not as steep as the short trail. The Mount Vaea Scenic Reserve is managed by the Department of Lands, Surveys, and Environment.

The tiny Robert Louis Stevenson Memorial Reserve (0.4 ha, 1 a.) encompasses the knoll just below the summit of Mt. Vaea. It is here that Robert Louis Stevenson and his wife are buried. Robert Louis Stevenson, locally known as Tusitala ("storyteller"), spent the last four years of his life in Sāmoa.

At the base of the mountain is the Robert Louis Stevenson Museum, the author's home during his time in Sāmoa. It was recently beautifully restored by the RLS Museum/Preservation Foundation to commemorate the centenary of Stevenson's death. Stevenson named his home Vailima, meaning "five waters" (referring to streams on the property). Today, the local award-winning Sāmoan beer shares the same name.

The Vailima Botanical Garden encompasses about 12 ha (30 a.) and protects a variety of species—native, introduced, and naturalized—representative of the flora of the archipelago and the Pacific basin. Although heavily damaged by recent hurricanes, the garden is gradually recovering. The garden is managed by the Department of Lands, Surveys, and Environment.

Biodiversity Conservation Sites

Launched by the South Pacific Regional Environmental Programme (SPREP), the Biodiversity Conservation Sites throughout the South Pacific have as their mission "to conserve the biodiversity of the conservation areas based on sustainable use of natural resources for the benefit of current and future generations." In many cases, income-generating activities linked to biodiversity conservation, such as ecotourism, are promoted and supported by SPREP. These activities encourage sustainable use and, by earning income for communities, reinforce community commitment to conservation.

Several Biodiversity Conservation Sites have been

identified for conservation, representing the minimum necessary to preserve the ecological diversity of the western islands of Sāmoa. None of these sites are completely secure from the threats of development associated with population growth and demand for local natural resources (e.g., timber). Nevertheless, the local government, SPREP, and local citizens' environmental groups are working together with villages to manage these ecologically important areas for conservation purposes.

Numerous essential conservation sites have been identified on the islands of 'Upolu and Savai'i. Visitors can show support for the villages managing their lands for conservation by visiting the following areas and possibly staying overnight:

• Uafato-Ti'avea Coastal Forest is one of the few areas left in the western islands of Sāmoa in which a relatively undisturbed band of rain forest habitats runs from the sea to the mountain interior. In particular, Uafato (14 sq. km) boasts an extensive area of mature coastal forest containing **ifilele** (*Intsia bijuga*) trees. The very dense wood of the **ifilele** is preferred by local carvers for kava bowls and other artifacts. The forest supports large numbers of flying foxes, and twenty-two bird species have been recorded in the Uafato forest, including the rare tooth-billed pigeon (*Didunculus strigirostris*), white-throated pigeon (*Columba vitiensis castaneicieps*), and **mao** (*Gymnomyza samoensis*). The regionally threatened coconut crab (*Birgus latro*), while scarce in most other areas in the western islands of Sāmoa, is more common in the conservation area and is often caught for food. Although essentially a rain forest reserve, this biodiversity conservation area also encompasses an adjacent marine area including spectacular coastal scenery and fringing reef. Its dramatic mountainous scenery is enhanced by numerous waterfalls cascading down the mountainsides. A number of small rivers with spectacular waterfalls flow through the conservation area. Uafato was also chosen as a biodiversity conservation area because of its cultural importance as an area of traditional resource use that remains viable and sustainable. It is also one of the first areas in the western islands of Sāmoa to be settled, perhaps 2,500 or 3,000 years ago, and it was here that local clay was used to make Lapita pottery, a cultural remnant that has helped archaeologists describe the migration and settling of the islands by the earliest Polynesian explorers. In addition, a number of historic sites associated with legends (**vavau**) occur here. The ancestral god Moso is said to have turned a flock of chickens into stone. These stone chickens can still be seen on the beach located on the western side of Uafato Bay.

This conservation area is located in one of the wettest areas of the island of ʻUpolu (600 cm of rainfall annually), on the northeast coast between the villages of Uafato and Tiʻavea. Management of this conservation area is coordinated by O le Siosiomaga Society, a local nongovernmental organization.

• Saanapu-Sataoa Mangrove and Coastal Rain Forest, situated on the south coast of ʻUpolu, is a beautiful estuary and the largest and best-quality mangrove forest remaining in the entire archipelago. It is the only mangrove conservation area in the western islands of Sāmoa, and is an extremely important fish and crab breeding site. Despite their importance, mangrove forests are under threat from land filling, pollution, and other habitat destruction.

Canoe rentals are available, and the clear, quiet waters of the mangrove swamp are an ideal place to paddle and observe wildlife. This conservation area is jointly managed by the Department of Lands, Surveys, and Environment and the local community.

• Nuʻutele-Nuʻulua-Fanuatapa-Namuʻa Islands, together known as the Aleipata Islands, are located off the eastern coast of 'Upolu. Long considered an important area worthy of protection, it has now been declared a marine park area. The islands themselves are important seabird nesting sites, and many are covered in relatively undisturbed lowland, coastal, and littoral rain forest vegetation.

• Vaoto Lowland, located on the south-central coast of Savai'i, near the Taga blowholes, is a lowland forest conserving high quality **tava** (*Pometia pinnata*) forest. This type of forest once occupied large expanses of the wetter lowlands of Savai'i.

• Aopo-Letui-Sasina Coastal Forest, located on the northwestern coast of Savai'i, is a large, high-quality area with important stands of **ifilele** (*Intsia bijuga*) lowland forest and scrub habitat.

• Lake Lanoto'o (**Vai Tuloto**), located high in the center of the island of 'Upolu, is the largest freshwater lake in the Sāmoan Archipelago. Although the largest, it covers only 16 ha (just under 39 a.), and fills an extinct volcanic crater whose depth is unknown. Although the lake itself is filled with nonnative goldfish and tilapia, it is bordered by a pandanus (**fala**) forest, and two of the rarest birds in the archipelago, the tooth-billed pigeon (*Didunculus strigirostris*, **manumea**) and **mao** (*Gymnomyza samoensis*, **ma'oma'o**), are known to breed in the montane rain forest surrounding Lake Lanoto'o.

CONVERSIONS

1 centimeter (cm) = .3937 inches (in.)
1 meter (m) = 3.2808 feet (ft.)
1 kilometer (km) = .6214 miles (mi.)
1 square kilometer (sq. km) = .3861 square miles
   (sq. mi.)
1 hectare (ha) = 2.471 acres (a.)
1 kilogram (kg) = 2.2046 pounds (lb.)

PARTS OF A FISH

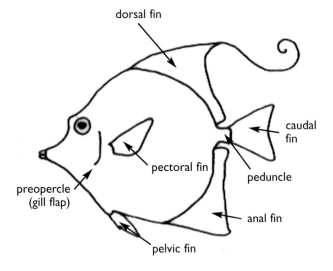

dorsal fin

caudal fin

pectoral fin

peduncle

preopercle
(gill flap)

anal fin

pelvic fin

GLOSSARY

algae: a nonflowering, stemless plant found in watery environments (e.g., seaweed, phytoplankton).

anterior: opposite, or facing away from, the center axis of the body.

archipelago: a large group of islands.

autotomy: the spontaneous discarding or dropping off of a body part when an organism is injured or under attack.

ava: (Sāmoan) a natural channel through the reef leading out to the ocean, often with swift currents and therefore dangerous.

benthic: living near or on the ocean bottom, irrespective of the depth.

biodiversity: the variety of plants and animals, including the genetic variation within all species.

carapace: the upper (dorsal) shell of a turtle; the **plastron** is the bottom (ventral) shell.

carnivorous: flesh-eating.

chlorophyll: the green pigment, typically found in plants but also in animals, which is used in the process of photosynthesis.

ciguatoxic: containing the toxin ciguatera, found in the flesh of some fish; ciguatera can cause severe illness in humans.

circumtropical: found living within the tropical zone, which is bounded by the Tropic of Capricorn and the Tropic of Cancer.

commensalism: a type of symbiosis; the interaction between species in which one species, the commensal, benefits from the other, sometimes called the host, while the host is not affected; compare with **mutualism** and **parasitism**.

coralline algae: a group of red seaweeds encrusted with lime that can function as reef builders; Rose Atoll is an algal reef composed of coralline algae.

cryptic: having coloration that renders an animal difficult to distinguish against its background.

diurnal: active during the day (compare with **nocturnal**).

dorsal: toward the upper surface of an organism or nearest the back; the opposite of **ventral**.

ecology: the branch of science that examines the relationship(s) between plants, animals, and their environment.

ectothermic: known commonly as "cold-blooded"; maintaining or regulating body temperature by behavioral means (e.g., basking in the sun or seeking shade); compare with **endothermic** and **exothermic**.

elongate: having more length (e.g., from head to tail) than width.

endangered (species): in imminent danger of extinction; compare with **threatened (species)**.

endemic (species): native; restricted to and naturally occurring in a specific geographic location and nowhere else in the world; compare with **introduced** and **native**.

endothermic: known commonly as "warm-blooded"; maintaining body temperature within certain temperature limits by means of internal mechanisms such as dilation of blood vessels, sweating, shivering, etc.; compare with **ectothermic** and **exothermic**.

epiphyte: a plant that grows nonparasitically on the surface of another plant.

eutrophication: the process by which high levels of nutrients encourage the growth of small plants, known as algae, which use up so much oxygen that nothing else grows.

exothermic: having a body temperature that varies according to the temperature of the surroundings; fish are exothermic (also called "poikilothermic"); compare with **ectothermic** and **endothermic**.

extinct: no longer existing.

extirpated: locally extinct, but possibly existing at other geographic locations.

fauna: all animal life.

feral: having reverted from a domestic state to a wild state following escape or release from captivity.

flora: all plant life.

frugivorous: fruit-eating.

herbivorous: plant-eating.

hermaphroditic: having both male (testes) and female (ovaries) sex glands at the same time.

herpetofauna: reptiles and amphibians as a group (even though they are not closely related).

introduced (species): not occurring naturally in a particular area, but instead brought to that area either knowingly or accidentally; nonnative; compare with **endemic** and **native**.

invertebrate: lacking a backbone or spinal column (e.g., land snails, octopuses, nudibranchs).

islet: tiny island.

laterally compressed: directed toward the sides.

marine: relating to the sea.

mimicry: the superficial resemblance of one species to another.

mutualism: a type of symbiosis; an interaction between members of two species in which both species benefit by the association; compare with **commensalism** and **parasitism**.

native (species): naturally occurring in a particular location; compare with **endemic** and **introduced**.

nematocyst: a term used to describe a sticky, poisonous, or otherwise entrapping microscopic structure used to capture food or for defense.

nocturnal: active at night; compare with **diurnal**.

nonnative (species): See **introduced**.

omnivorous: eating both plant and animal material.

ovate: egg-shaped, oval.

pala: (Sāmoan) swamp or wetland.

parasitism: a type of symbiosis; an interaction between species in which one organism (the parasite) lives in or on another (the host), from which it obtains food, shelter, or other requirements. Parasitism usually causes some harm to the host; such harm can range from almost none to severe illness resulting in death. Compare with **commensalism** and **mutualism**.

parthenogenesis: the development of an egg without the egg's undergoing fertilization; offspring are genetically identical to the single parent. Species containing only a single sex are referred to as "parthenogenic."

pelagic: oceanic; applied to organisms that live in open water; or, in birds, to those that spend the major portion of their lives at sea, coming to land only to breed.

phytoplankton: microscopic plants that live in fresh or salt water; compare with **zooplankton** and **plankton**.

plankton: passively floating or weakly swimming microscopic plant or animal life that lives in fresh or salt water. Compare with **phytoplankton** and **zooplankton**.

plastron: the bottom (ventral) shell of a turtle; the **carapace** is the upper (dorsal) shell.

proboscis: a protrusion, usually tubular, from the anterior (head) end of an animal.

proteandrous: beginning life as male, then maturing to

become female; compare with **protogynous**.

protogynous: beginning life as female, then maturing to become male; compare with **proteandrous**.

rhinopore: tentacle-like structure found on some nudibranch species.

symbiosis: the living together in close association of dissimilar organisms. Compare with **commensalism**, **mutualism**, and **parasitism**.

terrestrial: living or growing on land.

threatened (species) in imminent danger of becoming endangered; compare with **endangered (species)**.

ventral: closest to the ground or substrate (or farthest from the spine, notochord, etc.).

vertebrates: animals with a backbone or spinal column (e.g., mammals, birds, fishes, reptiles, amphibians, etc.).

zooplankton: microscopic animals that live in fresh or salt water; compare with **phytoplankton** and **plankton**.

zooxanthellae: microscopic, one-celled symbiotic algae that inhabit the internal tissues of corals and some other marine animals (e.g., giant clams).

SELECTED REFERENCES AND BIBLIOGRAPHY

Alder, G. H., and R. Dudley. 1994. Butterfly biogeography and endemism on tropical Pacific islands. *Biological Journal of the Linnean Society* 51: 151–62.

Allardice, R. W. 1991. *A simplified dictionary of modern Sāmoan.* Auckland: Polynesian Press.

Allen, G. 1996. *Marine life of the Pacific and Indian oceans.* Singapore: Periplus Editions (HK) Ltd.

Allen, G. R., and R. Steene. 1994. *Indo-Pacific coral reef field guide.* Singapore: Tropical Reef Research.

Amerson, A. B., Jr., W. A. Whistler, and T. D. Schwaner. 1982. *Wildlife and wildlife habitat of American Sāmoa.* Vol. 1, *Environment and ecology;* Vol. 2, *Accounts of flora and fauna.* Report submitted to U.S. Department of Interior, Fish and Wildlife Services, Washington, DC.

Balazs, G. H., P. Craig, B. R. Winton, and R. K. Miya. 1994. *Satellite telemetry of green turtles nesting at French Frigate Shoals, Hawaii, and Rose Atoll, American Sāmoa.* Proceedings of the Fourteenth Annual Symposium on Sea Turtle Biology and Conservation. NOAA Technical Memorandum NMFS-SEFSC-351: 184–87.

Bier, J. A. 1990. *Islands of Samoa: Tutuila, Manuʻa, ʻUpolu, Savaiʻi.* Honolulu: University of Hawaiʻi Press.

Birkeland, C., ed. 1997. *Life and death of coral reefs.* New York: Chapman and Hall.

Birkeland, C., R. H. Randall, A. L. Green, B. D. Smith, and S. Wilkens. 1996. Changes in the coral reef communities of Fagatele Bay National Marine Sanctuary and Tutuila Island, American Sāmoa, over the last two decades. Report to the National Oceanic and Atmospheric Administration. U. S. Department of Commerce.

Birkeland, C., R. H. Randall, A. L. Green, B. D. Smith, and S. Wilkens. 2001. Changes in the coral reef communities of Fagatele Bay National Marine Sanctuary and Tutuila Island, American Samoa, over the last two decades. Report to the National Oceanic and Atmospheric Administration. U. S. Department of Commerce.

Birkeland, C., R. Randall, R. Wass, R. Smith, B. Smith, and S. Wilkens. 1987. *Biological resource assessment of the Fagatele Bay National Marine Sanctuary.* NOAA Technical Memorandum NOS MEMD 3.

Bjorndal, K. A., ed. 1995. *Biology and conservation of sea turtles.* Washington, D.C.: Smithsonian Institution.

Burggren, W. W., and B. R. McMahon. 1988. *Biology of the land crabs.* Cambridge: Cambridge University Press.

Carlquist, S. 1974. *Island biology.* New York: Columbia University Press.

Cervino, J. M., R. L. Hayes, L. Kaufman, I. Nagelkerken, K. Patterson, J. W. Porter, G. W. Smith, and C. Quirolo. 1998. Coral disease. *Science* 28 (April 24): 485–660.

Clements, J. F. 1991. *Birds of the world: A checklist.* Vista, California: Ibis Publishing.

Colin, P. L., and C. Arneson. 1995. *Tropical Pacific invertebrates: A field guide to the marine invertebrates occurring on tropical Pacific coral reefs, seagrass beds, and mangroves.* Beverly Hills, California: Coral Reef Press.

Comstock, J. A. 1966. *Lepidoptera of American Sāmoa with particular reference to biology and ecology.* Pacific Insects Monograph 11. Honolulu: Bernice P. Bishop Museum.

Connor, R. C., and D. M. Peterson. 1994. *The lives of whales and dolphins.* American Museum of Natural History. New York: Henry Holt and Company, Inc.

Cook, R. P., and D. Vargo. 2000. Range extension of *Doleschallia tongana* (Nymphalidae) to the Samoan archipelago, with notes on its life history and ecology. *Journal of the Lepidopterists' Society* 54 (1): 33–35.

Cowie, R. H. 1992. Evolution and extinction of Partulidae, endemic Pacific Island land snails. *Phil. Trans. R. Soc. Lond.* 335: 167–91.

———. 1998. Catalog of the nonmarine snails and slugs of the

Sāmoan Islands. Bishop Museum Bulletin in Zoology 3. Honolulu: Bernice P. Bishop Museum.

Cox, P. A. 1983. Observations on the natural history of the Sāmoan bats. *Mammalia* 47(4): 519–23.

———. 1997. *Nafanua: Saving the Sāmoan rainforest.* New York: W.H. Freeman and Company.

Cox, P. A., T. Elmquist, E. Pierson, and W. Rainey. 1991. Flying foxes as strong interactors in South Pacific island ecosystems: a conservation hypothesis. *Conserv. Biol.* 5: 448–54.

Craig, P., ed. 1993. *American Sāmoa: Natural history and conservation topics.* Vol. 1. Department of Marine and Wildlife Resources, American Sāmoa Government.

Craig, P. ed. 1995. *American Sāmoa: Natural history and conservation topics.* Vol. 2. Department of Marine and Wildlife Resources, American Sāmoa Government.

Craig, P., T. E. Morrell, and K. So'oto. 1994. Subsistence harvest of birds, fruit bats, and other game in American Sāmoa, 1990–1991. *Pacific Science* 48(4): 344–52.

Craig, P., B. Ponwith, F. Aitaoto, and D. Hamm. 1993. The commercial, subsistence, and recreational fisheries of American Sāmoa. *Marine Fisheries Review* 55(2): 109–15.

Craig, P., P. Trail, and T. E. Morrell. 1994. The decline of fruit bats in American Sāmoa due to hurricanes and overhunting. *Biol. Cons.* 69: 261–66.

Cribb, P., and W. A. Whistler. 1996. *Orchids of Sāmoa.* The trustees of the Royal Botanic Gardens, Kew.

Durrell, G. 1990. The Ark's anniversary. Arcade Publishing. New York: Little Brown and Company.

Elmqvist, T., P. A. Cox, W.E. Rainey, and E.D. Pierson, eds. 1993. The rain forest and the flying foxes: An introduction to the rain forest preserves on Savai'i, Western Sāmoa. Salelologa P.O., Savai'i, Western Sāmoa: Fa'asao Savai'i, The Conservation Society.

Gill, B. J. 1993. The land reptiles of Western Sāmoa. *Journal of the Royal Society of New Zealand* 23(2): 79–89.

Gosliner, T. M, D. W. Behrens, and G. C. Williams. 1996. *Coral reef animals of the Indo-Pacific.* Monterey, California: Sea Challengers.

Grant, G. S., S. A. Banack, and P. Trail. 1994. Decline of the sheath-tailed bat, *Emballonura semicaudata* (Chiroptera: Emballonuridae) on American Sāmoa. *Micronesica* 27(1/2): 133–37.

Grant, G. S., P. W. Trail, and R. B. Clapp. 1994. First specimens of sooty shearwater, Newell's shearwater, and white-faced storm-petrel from American Sāmoa. *Notornis* 41: 215–17.

Green, A. 1996. *Status of the coral reefs of the Sāmoan Archipelago.* Department of Marine and Wildlife Resources, American Sāmoa Government.

Green, A. L., and P. Craig. 1996. Rose Atoll: a refuge for giant clams in American Sāmoa. Biological Report Series. Pago Pago, American Sāmoa: Department of Marine and Wildlife Resources, American Sāmoa Government.

Harrison, C. S. 1990. *Seabirds of Hawaii: natural history and conservation.* Comstock Publishing Assoc. Ithaca, New York: Cornell University Press.

Harrison, P. 1983. *Seabirds: an identification guide.* Boston: Houghton Mifflin Company.

————. 1996. *Seabirds of the world: A photographic guide.* Princeton, New Jersey: Princeton University Press.

Hayman, P., J. Marchant, and T. Prater. 1986. *Shorebirds: An identification guide to the waders of the world.* London: A & C Black.

Hopkins, G. H. E. 1927. Lepidoptera. Fasc. 1, Butterflies of Sāmoa and some neighbouring island-groups. In *Insects of Sāmoa and other Sāmoan terrestrial arthropoda.* Vol. 2, no. 3. London: The British Museum (Natural History).

Hunter, C. L., A. Friedlander, W. H. Magruder, and K. Z. Meier. 1993. Ofu reef survey: baseline assessment and recommendations for long-term monitoring of the proposed National Park, Ofu, American Sāmoa. Report to the National Park Service, Pago, Pago, American Sāmoa.

Ineich, I., and G. R. Zug. 1991. Nomenclatural status of *Emoia cyanura* (Lacertilia, Scincidae) populations in the Central Pacific. *Copeia* 1991(4): 1132–36.

IUCN Red List of Threatened Animals. 2000.Website: www.wcmc.org.uk:80/species/animals/animal_redlist.html

Jefferson, T. A., S. Leatherwood, and M. A. Webber. 1993. *FAO species*. Marine Mammals of the World Identification Guide. Rome: FA0.

Kami, K. S., and S. E. Miller. 1998. *Sāmoan insects and related arthropods: Checklist and bibliography.* Bishop Museum Technical Report No. 13. Honolulu: Bernice P. Bishop Museum.

Leatherwood, S., R. R. Reeves, and L. Foster. 1984. *Sierra Club handbook of whales and dolphins.* San Francisco: Sierra Club Books.

Lovegrove, T., B. Bell, and R. Hay. 1992. *The indigenous wildlife of Western Sāmoa: Impacts of Cyclone Val and a recovery and management strategy.* New Zealand Department of Conservation, on behalf of New Zealand Ministry of External Relations and Trade.

Lovell, E. R., and F. Toloa. 1994. *Palolo Deep National Marine Reserve: A survey, inventory and information report.* SPREP Report and Study Series No. 84. Apia, Western Sāmoa: South Pacific Regional Environment Programme.

McKeown, S. 1996. *A field guide to reptiles and amphibians in the Hawaiian Islands.* Los Osos, California: Diamond Head Publishing, Inc.

Madge, S., and H. Burn. 1988. *Waterfowl: an identification guide to the ducks, geese, and swans of the world.* Boston: Houghton Mifflin.

Miller, J. Y., and L. D. Miller. 1993. *The butterflies of the Tonga Islands and Niue, Cook Islands, with the descriptions of two new subspecies.* Bishop Museum Occasional Papers 34. Honolulu: Bernice P. Bishop Museum.

Miller, S. E. 1993. Final report on surveys of the arboreal and terrestrial snail fauna of American Sāmoa. Report submitted to U.S. Fish and Wildlife Service, Honolulu, Hawaii.

Milner, G. B. 1993. *Sāmoan dictionary.* Aotearoa, New Zealand: Polynesian Press,

Moors, P. J., ed. 1985. *Conservation of island birds: Case studies for the management of threatened island species.* International Council for Bird Preservation Technical Publication No. 3. Cambridge, England.

Morton, J., M. Richards, S. Mildner, and L. Bell. 1993. *The shore ecology of Upolu, Western Sāmoa.* 2d ed. Warkworth, New Zealand: University of Auckland.

Mundy, C. 1996. A quantitative survey of the corals of American Sāmoa. Report to the Department of Marine and Wildlife Resources, American Sāmoa Government.

Muse, C., and S. Muse. 1982. *The birds and birdlore of Sāmoa.* Walla Walla, Washington: Pioneer Press.

Myers, R. F. *Micronesian reef fish: A practical guide to the identification of the coral reef fishes of the tropical Central and Western Pacific.* Barrigada, Guam: Coral Graphics.

Parsons, M. 1991. *Butterflies of the Bulolo-Wau Valley.* Honolulu: Bishop Museum Press.

Perrine, D. 1997. *Mysteries of the sea.* Lincolnwood, Illinois: Publications International, Ltd.

Pierson, E. D., T. Elmqvist, W. E. Rainey, and P. A. Cox. 1996. Effects of tropical cyclonic storms on flying fox populations of the South Pacific Islands of Sāmoa. *Conservation Biology* 10(2): 438–51.

Pietson, E. D., and W. E. Rainey. 1992. The biology of flying foxes of the genus *Pteropus*: a review. *Pacific island flying foxes: Proceedings of an international conservation conference,* edited by D.E. Wilson and G.L Graham. U.S. Fish and Wildlife Service Biological Report 90 (23): 1–17.

Pratt, H. D., P. L. Bruner, and D. G. Berrett. 1987. *A field guide to the birds of Hawaii and the tropical Pacific.* Princeton, New Jersey: Princeton University Press.

Randall, J. E, G. R. Allen, and R. C. Steene. 1996. *Fishes of the Great Barrier Reef and Coral Sea.* 2d ed. Bathurst, Australia: Crawford House Press.

Rodes, B. K., and R. Odell. 1992. *A dictionary of environmental quotations*. New York: Simon and Schuster.

Rodgers, K. A., I. A. W. McAllan, C. Cantrell, and B. J. Ponwith. N.d. *Rose Atoll: An annotated bibliography*. Technical Reports of the Australian Museum, no. 9: 1–37.

Ryan, P. 1994. *The snorkeler's guide to the coral reef: From the Red Sea to the Pacific Ocean*. Honolulu: University of Hawai‘i Press.

Sale, P. F., ed. 1991. The ecology of fishes on coral reefs. San Diego, California: Academic Press, Inc.

Schwaner, T. D. 1979. Biogeography, community ecology and reproductive biology of the herpetofauna of the American Sāmoan [sic] islands. Ph.D. diss., University of Kansas.

Seto, N. W. H., and S. Conant. 1996. The effects of rat (*Rattus rattus*) predation on the reproductive success of the Bonin Petrels (*Pterodroma hypoleuca*) on Midway Atoll. *Colonial Waterbirds*. Bull. 19, no. 2: 171–185.

Sison, P. 1953. Further studies on the biology, ecology, and control of the black beetle of coconut *Orcytes rhinoceros* Linn. Proceedings of the Eighth Pacific Science Congress, 1953, Philippines.

South Pacific Regional Environmental Programme. Year of the coral reef fact sheet. Apia, Western Sāmoa.

Stone, C. P., and L. W. Pratt. 1994. Hawai‘i's plants and animals: Biological sketches of Hawai‘i Volcanoes National Park. Volcano, Hawai‘i: Hawai‘i Natural History Association.

Talbot, D., and D. Swaney. 1998. *Samoa: Independent and American Samoa*. 3d ed. Victoria, Australia: Lonely Planet.

Taule‘alo, T. 1993. *Western Sāmoa State of the Environment Report*. Apia, Western Sāmoa: South Pacific Regional Environmental Program.

Toone, B., D. Byers, R. Wirth, S. Ellis, and U. Seal. 1991. Conservation assessment and management plan for pigeons and doves. Report from a Workshop held 10–13

March 1993, San Diego, California. IUCN/SSC Captive Breeding Specialist Group, Birdlife International.

University of Hawai'i Cartographic Laboratory (1981). *Coastal Zone Management Atlas of American Samoa.* Honolulu: University of Hawai'i.

U.S. Fish and Wildlife Service. 1997. The impact of a ship grounding and associated fuel spill at Rose Atoll National Wildlife Refuge, American Sāmoa. Honolulu: Pacific Islands Ecoregion.

Volk, R. D., P. A. Knudsen, K. D. Kluge, and D. J. Herdrich. 1992. *Toward a territorial conservation strategy and the establishment of a conservation areas system for American Sāmoa.* Pago Pago, American Sāmoa: Le Vaomatua, Inc.

Walls, J. G., ed. 1982. *Encyclopedia of marine invertebrates.* Hong Kong: T.F.H. Publications, Inc., Ltd.

Wass, R. C. 1984. An annotated checklist of the fishes of Sāmoa. NOAA Technical Report. NMFS SSRF-781. U.S. Department of Commerce.

Whistler, W.A. 1994. *Botanical inventory of the proposed Tutuila and Ofu units of the National Park of American Sāmoa.* Technical Report 87. Cooperative Agreement CA8034-2-0001. Cooperative National Park Resources Study Unit. Honolulu: University of Hawai'i at Mānoa.

Wilson, R. W., and J. Q. Wilson. 1992. *Watching fishes: understanding coral reef fish behavior.* Pisces Books. Houston, Texas: Gulf Publishing Co.

Wingert, E. A. 1981. *Atlas of American Sāmoa.* Coastal Zone Management Program, Development Planning Office, American Sāmoa Government.

Zann, L. 1991. The inshore resources of Upolu, Western Sāmoa: coastal inventory and fisheries database. FAO/UNDP SAM/89/002 Field Report No. 5.

Zug, G. R. 1991. *Lizards of Fiji: natural history and systematics.* Bishop Museum Bulletin in Zoology 2. Honolulu: Bishop Museum Press.

————. 1993. Herpetology: An introductory biology of amphibians and reptiles. San Diego, California: Academic Press, Inc.

INDEX